Exercises *for* *the*
MICROBIOLOGY
LABORATORY

Burton E. Pierce
San Diego City College

Michael J. Leboffe
San Diego City College

Morton Publishing Company
925 W. Kenyon Ave., Unit 12
Englewood, Colorado 80110
http://www.morton-pub.com

Book Team

Publisher:	Douglas N. Morton
Biology Editor:	Chris Rogers
Production Manager:	Joanne Saliger
Typography:	Ash Street Typecrafters, Inc.
Cover Design:	Bob Schram, Bookends, Inc.

Preface

Exercises for the Microbiology Laboratory is an outgrowth and update of *Procedures Manual for the Microbiology Laboratory*. Whereas the *Procedures Manual* was tied directly to the first edition of the *Photographic Atlas for the Microbiology Laboratory* item for item, *Exercises* contains study units arranged in functional groups. The current arrangement is more likely to match the order in which the concepts will be introduced. Furthermore, the groupings offer functional continuity missing from the *Procedures Manual*.

The applications and theory (purpose and principle) of each exercise are given in the *Photographic Atlas for the Microbiology Laboratory* along with relevant photographs and illustrations not included in this book. We refer to it often. And, while each book may be used alone, their loose leaf design allows them to be combined into a comprehensive laboratory manual. We strongly recommend that you do this to get the "complete picture."

The specific order in which the exercises appear reflects our belief that simple and basic steps must precede complicated techniques which require greater skill. We, therefore, begin with laboratory safety followed by basic media preparation, microbial cultivation and growth characteristics, staining and microscopy, isolation techniques, and differential biochemical tests. As a reinforcement and practical application of these concepts, two exercises requiring identification of unknown bacterial species appear next. These are followed by quantitative techniques, medical, food and environmental microbiology, and hematology, immunology and serology. Finally, we offer some practice looking at and identifying the stages and various structures of eukaryotic microbes, the fungi, protozoa, and helminths. Recipes for all media, stains and reagents used appear in the appendices.

The basic design for each activity in *Exercises* is:

- page reference to the *Photographic Atlas for the Microbiology Laboratory*, 2nd Ed.
- materials list
- exercise protocol
- precautions
- references

A materials list is included for each exercise and is designed for a single student or lab group. Protocols provide step-by-step instructions for the performance of each exercise. Precautions are included to warn you of potential problems or areas of an exercise that need a little more care. Don't let their position near the end mislead you! Pay attention to these and you should meet with success in your lab work. In some cases, alternative procedures and more extensive explanations are also given in the Precautions. References are included for most procedures that will guide you in further study if you want more information about a particular test or technique.

It is our hope that *Exercises for the Microbiology Laboratory* helps to make your experience in Microbiology enjoyable and rewarding. We appreciate and encourage your comments.

Burton E. Pierce
Michael J. Leboffe

Acknowledgments

We remain indebted to all those who assisted and advised us during the development of the original *Procedures Manual for the Microbiology Laboratory* and also in its current incarnation, *Exercises for the Microbiology Laboratory*. An acknowledgment in a book can never repay your kindness and support, but it is a start!

James Bartley – San Diego City College
David Brady – San Diego City College
Deborah Durand – UCSD Department of Medicine
Thomas Lee – Scientific Instrument Company
William McClellan – Scientific Instrument Company
Let Negado – San Diego County Public Health Laboratory
Darla Newman – San Diego Mesa College
Dr. Margaret Polley – Calbiochem-Novabiochem Corporation
Debra Reed – San Diego City College
David Singer – San Diego City College
Robert Waddell – Scientific Instrument Company
Gary Wisehart – San Diego City College

Thanks also to those who used *Procedures* and offered valuable suggestions for its improvement. Your input was indispensable and resulted in its transformation into this work, one that is different enough to warrant a new title.

We are especially grateful to Natalie Cederquist of Hummingbird Graphics for her excellent illustrations. Thanks also to Doreen Cantelmo of Olympus America, for the microscope photograph and permission to reproduce it.

Joanne Saliger of Ash Street Typecrafters, Inc., Englewood, Colorado is most responsible for the overall appearance of this book as well as our others. Her pride and professionalism in producing attractive and quality products are greatly appreciated. We couldn't do it without you and your staff. In fact, a subtitle to this book could be, "Don't worry. Joanne will take care of it."

As always, we enjoy the support and encouragement of the people at Morton Publishing Company. Doug Morton, Christine Morton and Chris Rogers are the people we see and deal with personally, but we know that the whole Morton team is necessary for successfully completing a project of this sort. Thanks, and keep up the good work.

Writing a book (even a relatively small one) requires an incredible amount of time and effort on the part of the authors. Unfortunately, not all the sacrifices are made by the authors. Inevitably, the authors' loved ones are also affected. To our wives, Michele Pierce and Karen Leboffe, we publicly acknowledge your continued love and encouragement during difficult times you would rather have not gone through. It only makes us love you more. Thanks to both of you.

Contents

Safety and Laboratory Guidelines   vii

1   Fundamental Skills for the Microbiology Laboratory   1

Basic Growth Media

Exercise 1-1   Nutrient Agar and Nutrient Broth Preparation   2

Microbial Transfer Methods

Obtaining the Sample to be Transferred   5
Transferring to a Sterile Medium   11
Exercise 1-2   Aseptic Transfers   18

2   Cultivation and Growth of Bacterial Cultures   19

Gross Appearance of Bacterial Growth

Exercise 2-1   Bacterial Colony Morphology   20
Exercise 2-2   Growth Patterns in Broth   22

Aerotolerance

Exercise 2-3   Agar Deep Stabs   23

Anaerobic Growth

Exercise 2-4   Thioglycolate Broth   24
Exercise 2-5   Anaerobic Jar   25

3   Microscopy and Staining   27

Basic Light Microscopy

Guidelines for Use of the Light Microscope   28
Exercise 3-1   Examination of Prepared Microscope Slides   30
Exercise 3-2   Calibration of the Ocular Micrometer   31

Simple Bacteriological Stains

Exercise 3-3   Preparing a Bacterial Smear   34
Exercise 3-4   Simple Stain   36
Exercise 3-5   Negative Stain   38

Differential Stains

Exercise 3-6   Gram Stain   40
Exercise 3-7   Acid-Fast Stains   43
Exercise 3-8   Capsule Stain   45
Exercise 3-9   Spore Stain   46
Exercise 3-10   Flagella Stain   48
Exercise 3-11   Hanging Drop and Wet Mount Preparation   49

4   Isolation Techniques and Selective Media   51

Streaking for Microbial Isolation

Exercise 4-1   Streak Plate Method of Isolation   52

Selective Media for Isolation of Gram-Positive Cocci

Exercise 4-2   Blood Agar   54
Exercise 4-3   Mannitol Salt Agar (MSA)   56
Exercise 4-4   Phenylethyl Alcohol Agar (PEA)   57

Selective Media for Isolation of Gram-Negative Rods

Exercise 4-5   Desoxycholate (DOC) Agar   58
Exercise 4-6   Eosin Methylene Blue (EMB) Agar   59
Exercise 4-7   Hektoen Enteric (HE) Agar   60
Exercise 4-8   MacConkey Agar   61
Exercise 4-9   Xylose Lysine Desoxycholate (XLD) Agar   62

5   Differential Tests   63

Introduction to Energy Metabolism Tests

Exercise 5-1   Oxidation-Fermentation (OF) Medium   64

Tests Identifying Microbial Ability to Ferment

Exercise 5-2   Fermentation Tests (Phenol Red Broth)   66
Exercise 5-3   Fermentation Tests (Purple Broth)   68
Exercise 5-4   Methyl Red and Voges-Proskauer (MR/VP) Tests   69

Tests Identifying Microbial Ability to Respire

Exercise 5-5   Catalase Test   71
Exercise 5-6   Oxidase Test   73
Exercise 5-7   Nitrate Reduction Test   74

Utilization Media

Exercise 5-8   Citrate Utilization Test   76

Decarboxylation and Deamination Tests

Exercise 5-9   Decarboxylase Test   77
Exercise 5-10   Phenylalanine Deaminase Test   79

Tests Detecting the Presence of Hydrolytic Enzymes

Exercise 5-11   Bile Esculin Test   80
Exercise 5-12   Urease Tests (Agar)   81
Exercise 5-13   Urease Tests (Broth)   82
Exercise 5-14   Casease Test   83
Exercise 5-15   DNase Test   84
Exercise 5-16   Gelatin Liquefaction Test   85
Exercise 5-17   Lipase Tests   86
Exercise 5-18   Starch Hydrolysis Test   87

Combination Differential Media

Exercise 5-19 Kligler's Iron Agar   88
Exercise 5-20 Litmus Milk Medium   90
Exercise 5-21 Lysine Iron Agar (LIA)   91
Exercise 5-22 SIM Medium (Sulfur Reduction Test,
            Indole Production, Motility)   92
Exercise 5-23 Triple Sugar Iron (TSI) Agar   93

Antibiotic Susceptibility Testing

Exercise 5-24 Bacitracin Susceptibility Test   94

Other Differential Tests

Exercise 5-25 Coagulase Tests (Slide)   96
Exercise 5-26 Coagulase Tests (Tube)   97
Exercise 5-27 Motility Test   98

6 Determination of Bacterial Unknowns: Two Projects   99

Exercise 6-1 Morphological Unknown   100
Exercise 6-2 Bacterial Unknowns Project   104

7 Quantitative Techniques   111

Exercise 7-1 Viable (Plate) Count   112
Exercise 7-2 Direct Count   115
Exercise 7-3 Plaque Assay for Determination
            of Phage Titer   116
Exercise 7-4 Semiquantitative Streak
            of a Urine Specimen   119
Exercise 7-5 Bacterial Growth in a Closed System   120

8 Medical, Food and Environmental Microbiology   123

Producing and Detecting Mutagens

Exercise 8-1 Ultraviolet Radiation: Its Characteristics
            and Effects on Bacterial Cells   124
Exercise 8-2 Ames Test   125

Determining the Antibiotic of Choice for Treatment

Exercise 8-3 Antibiotic Sensitivity — Kirby-Bauer Test   127

Tests Identifying Bacterial Contamination in Samples

Exercise 8-4 Membrane Filter Technique   129
Exercise 8-5 MPN (Most Probable Number) Method
            for Total Coliform Determination   131
Exercise 8-6 Methylene Blue Reductase Test   133
Exercise 8-7 Snyder Test   134

9 Hematology, Immunology, and Serology   135

Hematology and Immunology

Exercise 9-1 Differential Blood Cell Count   136
Exercise 9-2 Organs and Cells
            of the Immune System   138

Simple Serological Reactions

Exercise 9-3 Precipitation Reactions —
            Precipitin Ring   139
Exercise 9-4 Precipitation Reactions —
            Double-Gel Immunodiffusion   140
Exercise 9-5 Agglutination Reactions —
            Slide Agglutination   142
Exercise 9-6 Agglutination Reactions —
            Blood Typing   143

10 Eukaryotic Microbes   145

Exercise 10-1 Fungi   146
Exercise 10-2 Protozoans   150
Exercise 10-3 Helminth Parasites   155

Appendices

A Recipes for Media Used in this Manual   161

B Recipes for Stains and Reagents Used in this Manual   173

Safety and Laboratory Guidelines

Microbiology lab can be an interesting and exciting experience, but there are also potential hazards of which you should be aware. Improper handling of chemicals, equipment and/or microbial cultures is a dangerous practice and can result in injury or infection. Listed below are some general safety rules which must be followed in order to reduce the chance of injury or infection to you and to others. Please follow these and any other safety guidelines required by your college.

STUDENT CONDUCT

1. To reduce the risk of infection, do not smoke, eat, drink or bring food or drinks into the laboratory room — even if lab work is not being done.

2. Wash your hands *thoroughly* with soap and water before leaving the laboratory each day.

3. Come to lab prepared for that day's work. Lab time is precious. Besides, figuring out what to do as you go is an undertaking designed to produce confusion and accidents.

4. Do not remove any organisms or chemicals from the laboratory.

BASIC LABORATORY SAFETY

1. Wear protective clothing (*i.e.*, a lab coat) in the laboratory when handling microbes. It should be removed prior to leaving the lab and autoclaved regularly.

2. Wear eye protection whenever heating chemicals.

3. Turn off your Bunsen burner when not in use. Not only is it a fire and safety hazard, but an unnecessary source of heat in the room.

4. Tie long hair back. It is a potential source of contamination as well as a likely target for fire.

5. If you are feeling ill (for whatever reason) do not work with live microbes. There are other ways you may contribute to your lab group (*i.e.*, record data, fill out culture labels, retrieve equipment, *etc.*).

6. If you are pregnant or are taking immunosuppressant drugs, please see the instructor. It may be in your best *long*-term interests to postpone taking this class.

7. Wear disposable latex gloves while staining microbes and handling blood products (*i.e.*, plasma, serum, antiserum, or whole blood). Handling blood can be hazardous even with gloves, therefore, consult your instructor before attempting to work with any blood products.

8. Use an antiseptic (*e.g.*, Betadine) on your skin if it is exposed to a spill containing microorganisms. Your instructor will tell you which antiseptic you will be using.

9. Never pipette by mouth. Always use mechanical pipettors.

10. Dispose of uncontaminated broken glass in a "sharps" container.

11. Use a fume hood to perform any work which involves highly volatile chemicals or stains which need to be heated.

12. Find the first aid kit and make a mental note of its location.

13. Find the fire blanket and fire extinguisher, note their location and make a plan for how to get them in an emergency.

14. Find the eye wash basin, learn how to operate it and remember its location.

REDUCING CONTAMINATION OF SELF, OTHERS, CULTURES, AND THE ENVIRONMENT

1. Wipe the desk top with a disinfectant (*e.g.*, Amphyl) before *and* after each lab period in which live organisms are used. Appropriate disinfectant will be supplied.

2. Never lay culture tubes down on the table; they should always remain upright in a tube holder. Even solid media tubes contain moisture or condensation which may leak out and contaminate the work surface, your hands, or other cultures.

3. Cover any culture spills with paper towels. Immediately soak the towels with disinfectant and allow

them to stand for 20 minutes. Report the spill to your instructor. When finished, place the towels in the container designated for autoclaving.

4. Place books and papers other than what is essential for the lab period under the desk. A cluttered lab table is an invitation for an accident, an accident that may contaminate your expensive school supplies.

5. Place a disinfectant-soaked towel on the work area when pipetting. This reduces contamination and possible aerosols if a drop escapes from the pipette and hits the tabletop.

DISPOSAL OF CONTAMINATED MATERIALS

1. Dispose of plate cultures (if plastic Petri dishes are used) and other contaminated nonreusable items in the appropriate container to be sterilized (*e.g.,* the autoclave bag) when you are finished with them.

2. Remove all labels from tube cultures or contaminated reusable items and place them in the container designated for autoclaving.

3. Dispose of all blood product samples as well as disposable latex gloves in the container designated for autoclaving.

4. Place used microscope slides of bacteria in a container designated for autoclaving or soak them in disinfectant solution for at least 20 minutes before cleaning them.

5. Place contaminated broken glass in the container designated for autoclaving. After sterilization, dispose of the glass in a "sharps" container or specialized broken glass container.

REFERENCES

Barkley, W. Emmett and John H. Richardson. 1994. Chapter 29 in *Methods for General and Molecular Bacteriology,* edited by Philipp Gerhardt, R.G.E. Murray, Willis A. Wood, and Noel R. Krieg. American Society for Microbiology, Washington, D.C.

Collins, C.H., Patricia M. Lyne and J.M. Grange. 1995. Chapters 1 and 4 in *Collins and Lyne's Microbiological Methods, 7th Ed.* Butterworth-Heineman, Oxford.

Darlow, H. M. 1969. Chapter VI in *Methods in Microbiology, Volume 1,* edited by J. R. Norris and D. W. Ribbins. Academic Press, Ltd., London.

Fleming, Diane O. 1995. *Laboratory Safety - Principles and Practices, 2nd Ed.,* edited by Diane O. Fleming, John H. Richardson, Jerry J. Tulis and Donald Vesley. American Society for Microbiology, Washington, D.C.

Koneman, Elmer W., Stephen D. Allen, William M. Janda, Paul C. Schreckenberger, and Washington C. Winn, Jr. 1997. *Color Atlas and Textbook of Diagnostic Microbiology, 5th Ed.* Lippincott-Raven Publishers, Philadelphia and New York.

Power, David A. and Peggy J. McCuen. 1988. Pages 2 and 3 in *Manual of BBL® Products and Laboratory Procedures, 6th Ed.* Becton Dickinson Microbiology Systems, Cockeysville, MD.

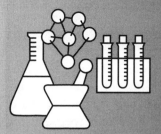

Fundamental Skills for the Microbiology Laboratory

*B*acteria and fungi are cultured and maintained on (or in) solid and liquid substances called *media*. Preparation of these media involves weighing ingredients, measuring liquid volumes, calculating proportions, handling basic laboratory glassware, and operating a pH meter and an autoclave. In this section, you will learn and practice these fundamental skills by preparing a few simple growth media. When you have completed the exercises you will be able to prepare virtually any medium if given the recipe. All of the recipes of media used in this manual are located in the Appendix.

A second fundamental skill necessary for any microbiologist is the ability to transfer bacterial cells from one place to another without contaminating the original culture, the new medium, or the environment (including the microbiologist). This *aseptic* (sterile) transfer technique is required for virtually all procedures in which living microbes are handled, including isolations, staining, and differential testing. The second part of this section, *Aseptic Transfers,* is devoted to descriptions of how to transfer bacterial samples from one medium to another.

BASIC GROWTH MEDIA

Microbial cultivation requires specialized growth media. These media may be prepared from scratch, or they may be made by rehydrating commercially available powdered media. Media routinely encountered in the microbiology laboratory range from the widely-used, general-purpose growth media, to the more specific selective and differential media used in identification of microbes. This unit will teach you how to prepare simple general growth media. (Preparation instructions for more complex media are located in Appendix A.)

Nutrient agar and nutrient broth are common media used for maintaining bacterial cultures. They are formulated to meet the diverse nutrient requirements of a majority of bacteria. As such, they are made from sources that supply carbon and nitrogen in a variety of forms — amino acids, purines, pyrimidines, mono- to polysaccharides and various lipids. Generally, these are provided in digests of plant material (phytone) or animal material (peptone and others). Since the exact composition and amounts of the carbon and nitrogen in these ingredients is unknown, general growth media are considered to be *undefined*.

While in most classes (due to limited time), media are prepared by a laboratory technician, it is instructive for novice microbiologists to at least gain exposure to what is involved in media preparation. Your instructor will provide specific instructions on how to execute this exercise using your particular laboratory equipment.

Exercise 1–1 — Nutrient Agar and Nutrient Broth Preparation

MATERIALS

Two-liter Erlenmeyer flasks
Three or four 500 mL Erlenmeyer flasks and covers (can be aluminum foil)
Stirring hotplate
Magnetic stir bars
All ingredients listed below in the recipes
Sterile Petri dishes
Test tubes and caps
Balance
Weighing paper or boats
Spatulas

RECIPES

Nutrient Agar

Beef extract	3.0 g
Peptone	5.0 g
Agar	15.0 g
Distilled or deionized water	1.0 L

final pH = 6.8 ± 0.2 at 25°C

Nutrient Broth

Beef extract	3.0 g
Peptone	5.0 g
Distilled or deionized water	1.0 L

final pH = 6.8 ± 0.2 at 25°C

MEDIA PREPARATION

Nutrient Agar Tubes

1. Suspend the ingredients in one liter of distilled or deionized water, mix well, and boil until fully dissolved.

2. Dispense 10 mL portions into test tubes and cap loosely.

3. Autoclave for 15 minutes at 121°C to sterilize the medium.

4. Cool to room temperature with the tubes in an upright position for agar deep tubes. Cool with the tubes on an angle for agar slants.

Nutrient Agar Plates

1. Suspend the ingredients in one liter of distilled or deionized water, mix well, and boil until fully dissolved.

2. Divide into three or four 500 mL flasks for pouring. (Smaller flasks are easier to handle when pouring plates.)

3. Cover the containers and autoclave for 15 minutes at 121°C to sterilize the medium.

4. Remove the sterile agar from the autoclave and allow it to cool to 50°C.

5. Dispense approximately 15 mL into sterile Petri plates. Gently swirl to get the agar to completely cover the base. Allow the agar to cool before moving the plates.

6. Invert the plates and store them on a countertop or in the incubator for 24 hours to allow them to dry slightly before inoculating.

Nutrient Broth

1. Suspend the ingredients in one liter of distilled or deionized water. Agitate and heat slightly (if necessary) to dissolve completely.

2. Dispense 7.0 mL portions into test tubes and cap loosely.

3. Autoclave for 15 minutes at 121°C to sterilize the medium.

PRECAUTIONS

⚠ To minimize contamination while preparing media, clean the work surface, turn off all fans and close any doors that might allow excessive air movements.

⚠ Shield the Petri dish with its lid while you pour agar to reduce the chance of introducing airborne contaminants.

REFERENCES

DIFCO Laboratories. 1984. Pages 619 and 622 in *DIFCO Manual*, *10th Ed.*, DIFCO Laboratories, Detroit, MI.

Power, David A. and Peggy J. McCuen. 1988. Pages 214 and 215 in *Manual of BBL® Products and Laboratory Procedures*, *6th Ed.*, Becton Dickinson Microbiology Systems, Cockeysville, MD.

MICROBIAL TRANSFER METHODS

As a microbiology student, you will be required to collect a specimen and then transfer it from its source to a sterile medium. Proper transfer technique is performed *aseptically* (*i.e.*, without contamination of the culture, the sterile medium or the surroundings). These transfers are basic to most work you will be doing this semester, so practice them until you become proficient.

To prevent contamination of the sample, inoculating instruments (Fig. 1-1) are sterilized prior to use. Instruments such as Pasteur pipettes, cotton applicators, and serological or Mohr pipettes are sterilized by autoclaving long before use and allowed to dry and/or cool. Inoculating loops and needles, on the other hand, are sterilized immediately before use in an incinerator or Bunsen burner flame. As an extra precaution, the lips of tubes or flasks containing cultures or media are incinerated at the time of transfer.

There are two basic stages of transfers: 1) obtaining the sample to be transferred, and 2) transferring to the sterile culture medium. Aseptic transfers are not difficult. However, a little preparation will make whichever one you use go much more smoothly. You need to know where the sample is coming from, the type of transfer instrument to be used, and the sample's destination before you begin. Until transfers become second nature to you, we recommend that you pull the appropriate pages for obtaining and transferring the sample and keep them in a visible place.

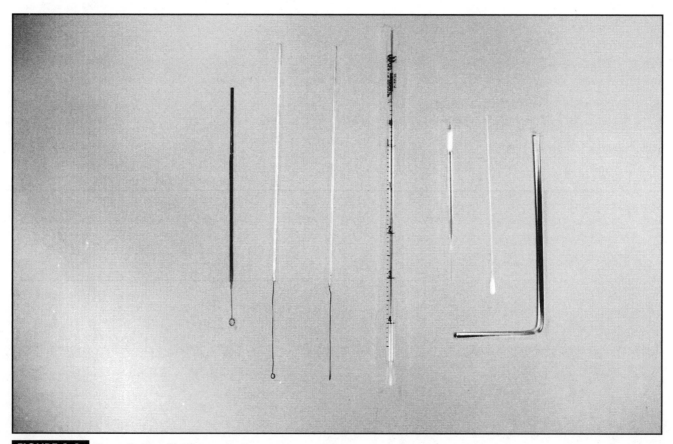

FIGURE 1-1 **Inoculating Tools**
Any of several different instruments may be used to transfer a microbial sample, the choice of which depends on the sample source, its destination, and any special requirements imposed by the specific protocol. Shown here are several examples of transfer instruments. From left to right: disposable inoculating loop, nichrome inoculating loop, inoculating needle, serological pipette, Pasteur pipette, cotton swab and glass spreading rod.

Obtaining the Sample to be Transferred

TRANSFERS WITH A STERILE COTTON SWAB

A cotton swab is usually used to obtain a sample from a patient or an environmental source, and occasionally from a culture grown in the laboratory. Sterile swabs may be dry or they may be in sterile water, depending on the sample source. In either case, care must be taken not to contaminate the swab by touching other surfaces with it. Examples are illustrated in Figures 1-2a and 1-2b, but your instructor may provide specific instructions on sample collection from other sources with a swab.

FIGURE 1-2a Sampling a Patient's Throat

A tongue depressor and swab prepared in sterile water may be used together to obtain a sample from the pharynx (throat). Touching other parts of the oral cavity is likely to cause contamination. Avoid touching the soft palate or a gag reflex may be initiated! The sample should then be transferred from the swab to a growth medium as quickly as possible. Your instructor may provide instructions on sampling other body regions.

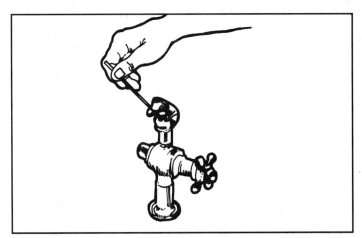

FIGURE 1-2b Sampling an Environmental Source

A swab prepared in sterile water may be used to obtain a sample from an environmental source. As with the throat culture, be careful to touch only the area to be sampled and transfer to the growth medium as soon as possible.

TRANSFERS FROM AN AGAR PLATE USING AN INOCULATING LOOP OR NEEDLE

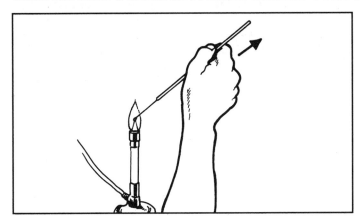

FIGURE 1-3a Flame the Loop/Needle

[Note: Since loops and needles are handled in the same way, we refer only to a loop in the following instructions for ease of reading.] Sterilize the loop by incinerating it in the Bunsen burner flame. Hold the handle like a pencil in your dominant hand and relax! Pass it through the tip of the inner cone of flame (the hottest part) holding it at an angle with the loop end pointing downward. Begin flaming about 2 cm up the handle, then proceed down the wire by pulling the loop backwards through the flame until the entire wire has become red-hot. Flaming in this direction limits aerosol production by allowing the tip to heat up more slowly than if it were thrust into the flame immediately.

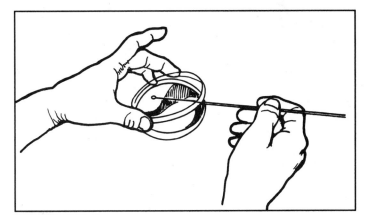

FIGURE 1-3b Use the Lid as a Shield

Lift the lid of the agar plate, but continue to use it as a cover to prevent contamination from above. Touch the loop to an uninoculated portion of the plate to cool it. (Placing a hot wire on growth may cause spattering of the growth and create aerosols.) Obtain a small amount of bacterial growth by gently touching a colony with the wire tip. Holding the transfer instrument very still to avoid creating aerosols, carefully remove it and replace the lid. What you do next depends on the medium to which you are transferring the growth. Please continue with the appropriate inoculation section.

TRANSFERS FROM AN AGAR SLANT USING AN INOCULATING LOOP OR NEEDLE

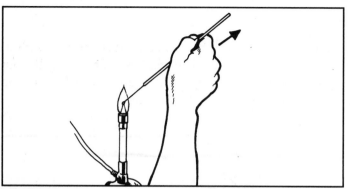

FIGURE 1-4a **Flame the Loop/Needle**
[**Note**: *Since loops and needles are handled in the same way, we refer only to a loop in the following instructions for ease of reading.*] Sterilize the loop by incinerating it in the Bunsen burner flame. Hold the handle like a pencil in your dominant hand and relax! Pass it through the tip of the inner cone of flame (the hottest part) holding it at an angle with the loop end pointing downward. Begin flaming about 2 cm up the handle, then proceed down the wire by pulling the loop backwards through the flame until the entire wire has become red-hot. Flaming in this direction limits aerosol production by allowing the tip to heat up more slowly than if it were thrust into the flame immediately.

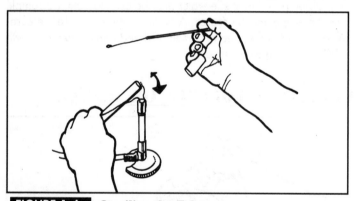

FIGURE 1-4c **Sterilize the Tube**
Pass the lip of the tube quickly through the flame two or three times to sterilize the glass and the surrounding air. The tube should be held on an angle to prevent contamination from above. Keep your loop hand still.

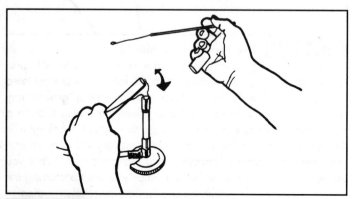

FIGURE 1-4e **Sterilize the Tube Again**
Flame the tube lip as before. Keep your loop hand still.

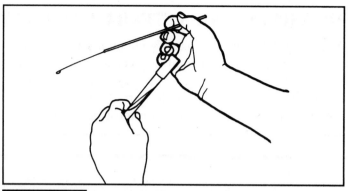

FIGURE 1-4b **Hold the Cap in Your Pinkie Finger**
While keeping your loop hand still, bring the culture tube towards it. Use the pinkie finger of your loop hand to remove and hold its cap. (The cap should be loosened prior to the transfer, especially if it's a screw top cap.)

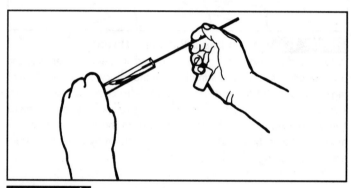

FIGURE 1-4d **Harvest the Growth**
With the agar surface facing upward, hold the open tube at an angle to prevent aerial contamination. Holding the loop hand still, move the tube up the wire until the wire tip is over the desired growth. Touch the loop to the growth and obtain the smallest visible mass of bacteria. Then, holding the loop hand still, *remove the tube* from the wire. Be especially careful when removing the tube not to catch the loop tip on the tube lip. This springing action of the loop creates bacterial aerosols.

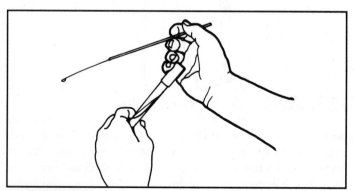

FIGURE 1-4f **Replace the Cap**
Keeping the loop hand still (remember, it has growth on it), move the tube to replace its cap. The cap at this point doesn't need to be on firmly — just enough to cover the tube. What you do next depends on the medium to which you are transferring the growth. Please continue with the appropriate inoculation section.

TRANSFERS FROM A BROTH CULTURE USING AN INOCULATING LOOP OR NEEDLE

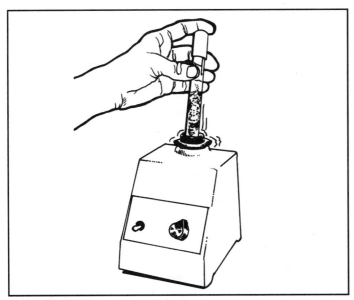

FIGURE 1-5a   **Suspend the Bacteria in the Broth (One Method)**

Growth may be suspended in the broth with a vortex mixer. Be sure not to mix so vigorously that broth gets into the cap or that you lose control of the tube. It's best to start slowly, then gently increase the speed until the tip of the vortex reaches the bottom of the tube.

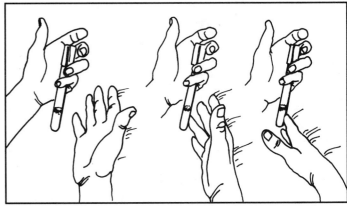

FIGURE 1-5b   **Suspend the Bacteria in the Broth (Another Method)**

The broth may also be agitated by drumming your fingers along the length of the tube several times. Be careful not to splash the broth into the cap or lose control of the tube.

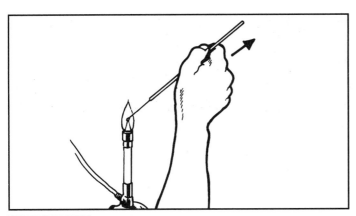

FIGURE 1-5c   **Flame the Loop**

[**Note**: *Since loops and needles are handled in the same way, we refer only to a loop in the following instructions for ease of reading.*] Sterilize the loop by incinerating it in the Bunsen burner flame. Hold the handle like a pencil in your dominant hand and relax! Pass it through the tip of the inner cone of flame (the hottest part) holding it at an angle with the loop end pointing downward. Begin flaming about 2 cm up the handle, then proceed down the wire by pulling the loop backwards through the flame until the entire wire has become red-hot. Flaming in this direction limits aerosol production by allowing the tip to heat up more slowly than if it were thrust into the flame immediately.

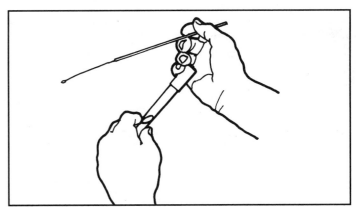

FIGURE 1-5d   **Hold the Cap in Your Pinkie Finger**

While keeping your loop hand still, bring the culture tube towards it. Use the pinkie finger of your loop hand to remove and hold its cap. (The cap should be loosened prior to the transfer, especially if it's a screw top cap.)

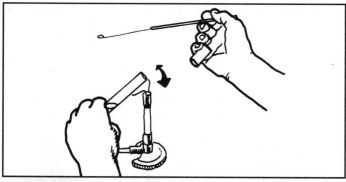

FIGURE 1-5e   Sterilize the Tube
Pass the lip of the tube quickly through the flame two or three times to sterilize the glass and the surrounding air. The tube should be held on an angle to prevent contamination from above. Keep your loop hand still.

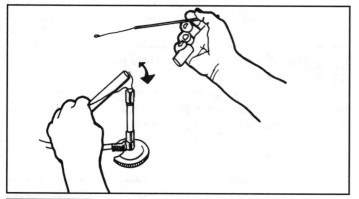

FIGURE 1-5g   Sterilize the Tube Again
Flame the tube lip as before. Keep your loop hand still.

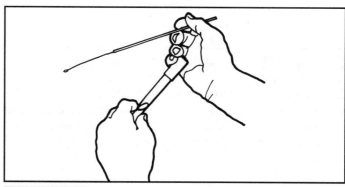

FIGURE 1-5f   Harvest the Growth
Hold the open tube at an angle to prevent aerial contamination. Holding the loop hand still, move the tube up the wire until the tip is in the broth. Continuing to hold the loop hand still, *remove the tube* from the wire. Be especially careful when removing the tube not to catch the loop tip on the tube lip. This springing action of the loop creates bacterial aerosols.

FIGURE 1-5h   Replace the Cap
Keeping the loop hand still (remember, it has growth on it), move the tube to replace its cap. The cap at this point doesn't need to be on firmly — just enough to cover the tube. What you do next depends on the medium to which you are transferring the growth. Please continue with the appropriate inoculation section.

PIPETTES AND PIPETTORS

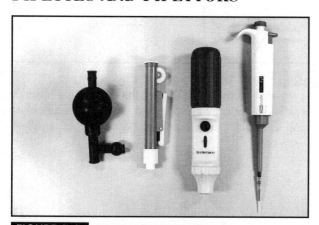

FIGURE 1-6   Mechanical Pipettors
Mouth pipetting is dangerous and has been replaced by mechanical pipettors. Several examples are shown here, each with its own method of operation. Your instructor will show you how to properly use the style of pipettor available in your lab. From left to right: a pipette bulb, a plastic pump, a pipette filler/dispenser, and a micropipettor.

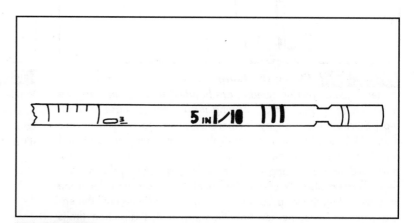

FIGURE 1-7a   Pipette Calibration
Read the pipette calibration. The numbers indicate the pipette's *total volume* and its *smallest calibrated increments*. This is a 5.0 mL pipette divided into 0.1 mL increments

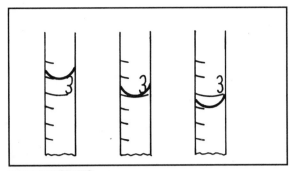

FIGURE 1-7b   **Read the Base of the Meniscus**

When reading volumes, use the base of the meniscus. The volume in the center pipette is read at exactly 3.0 mL because the meniscus is resting on the line. The left pipette is read at 2.9 mL and the right pipette is read as 3.1 mL (0 is always at the pipette's top). Although the difference in volume between these three pipettes may seem negligible (1 part in 30, a 3% error), it may be enough to introduce substantial error into your work.

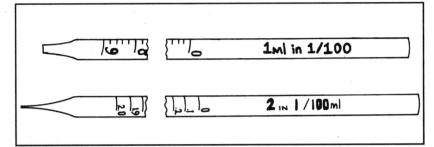

FIGURE 1-7c   **Two Types of Pipettes**

Two pipette styles are used in microbiology. These are the *serological pipette (above)* and the *Mohr pipette (below)*. A serological pipette is calibrated *to deliver* (TD) its volume by completely draining it and blowing out the last drop. The tip of a Mohr pipette is not graduated, so fluid flow must be stopped at a calibration line. Stopping the fluid beyond the last line on a Mohr pipette results in an unknown volume being dispensed. (If pipetting a bacterial culture, be careful not to allow any to drop from the pipette before disposing of it in the autoclave container. Clean up any spills.) In either case, volumes are read at the bottom of the meniscus of fluid.

TRANSFERS FROM A BROTH CULTURE USING A PIPETTE

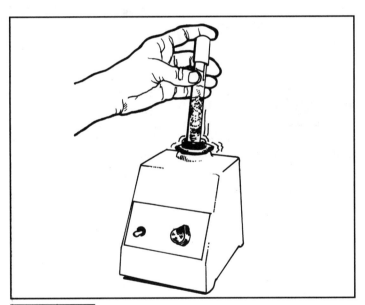

FIGURE 1-8a   **Suspend the Bacteria in the Broth (One Method)**

Growth may be suspended in the broth with a vortex mixer. Be sure not to mix so vigorously that broth gets into the cap or that you lose control of the tube. It's best to start slowly, then gently increase the speed until the tip of the vortex reaches the bottom of the tube.

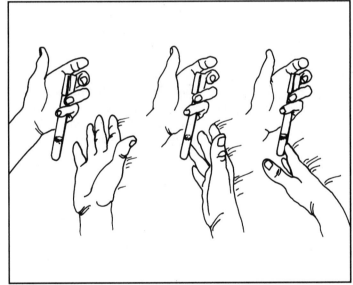

FIGURE 1-8b   **Suspend the Bacteria in the Broth (Another Method)**

The broth may also be agitated by drumming your fingers along the length of the tube several times. Be careful not to splash the broth into the cap or lose control of the tube.

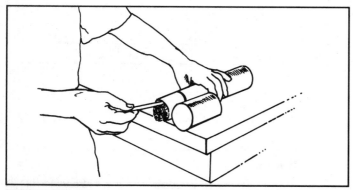

FIGURE 1-8c   **Get the Sterile Pipette**

Pipettes are sterilized in metal canisters or packages and are stored in groups of a single size. *Be sure you know what volume your pipette will deliver.* Set the canister at the table edge and remove its lid. (Canisters should not be stored in an upright position as they may fall over and break the pipettes or become contaminated.) If using pipettes in a package, open the end *opposite the tips.* Grasp *one pipette only* and remove it.

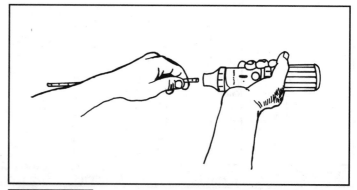

FIGURE 1-8d   **Assemble the Pipette**

Carefully insert the pipette into the mechanical pipettor. It's best to grasp the pipette near the end with your finger tips. This gives you more control and reduces the chances that you will break the pipette and cut your hand. *Do not touch* any part of the pipette that will contact the specimen or the medium or you risk introducing a contaminant. Also, do not lay the pipette on the table top while you continue.

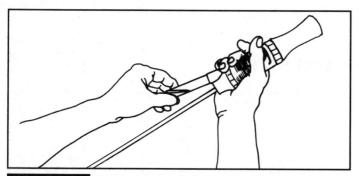

FIGURE 1-8e   **Hold the Cap in Your Pinkie Finger**

While keeping the pipette hand still, bring the culture tube towards it. Use your pinkie finger to remove and hold its cap. (The cap should be loosened prior to the transfer, especially if it's a screw top cap.)

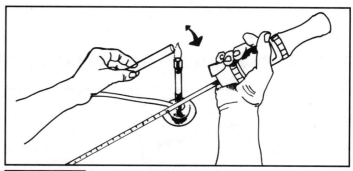

FIGURE 1-8f   **Sterilize the Tube**

Pass the lip of the tube quickly through the flame two or three times to sterilize the glass and the surrounding air. The tube should be held on an angle to prevent contamination from above. Keep your pipette hand still.

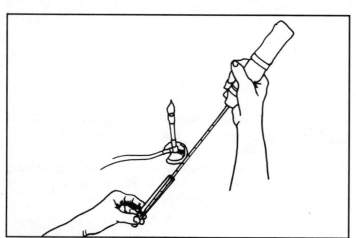

FIGURE 1-8g   **Remove the Desired Volume**

Insert the pipette and withdraw the appropriate volume. Bring the pipette to a vertical position briefly to accurately read the pipette. (Remember: the volumes in the pipette are correct only if the meniscus of the fluid inside is resting *on* the line, not below it.) Then carefully remove the pipette.

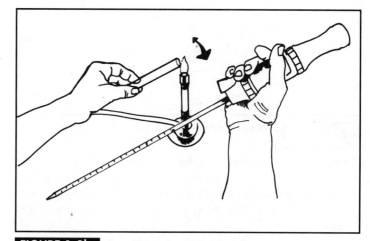

FIGURE 1-8h   **Sterilize the Tube Again**

Flame the tube lip as before. Keep your pipette hand still.

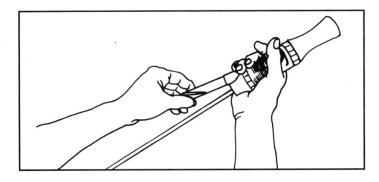

FIGURE 1-8i   **Replace the Cap**
Keeping the pipette hand still (remember, it contains fluid with microbes in it), move the tube to replace its cap. The cap at this point doesn't need to be on firmly — just enough to cover the tube. What you do next depends on the medium to which you are transferring the growth. Please continue with the appropriate inoculation section.

Transferring to a Sterile Medium

INOCULATION OF AGAR PLATES USING A COTTON SWAB

Since the cotton swab was probably used to obtain a specimen containing a mixed culture of microbes, agar plates are the typical medium inoculated. Depending on your purposes, you may inoculate the agar surface in a couple of different ways. A zigzag inoculation is shown in Figure 1-9a. Figure 1-9b shows inoculation with the swab in preparation for a streak plate.

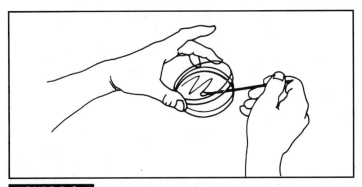

FIGURE 1-9a   **Zigzag Inoculation**
This inoculation pattern is usually performed when the sample does not have a high cell density or with pure cultures when isolation is not necessary. Hold the swab comfortably in one hand and lift the lid of the Petri dish with the other. Use the lid as a shield to protect the agar from aerial contamination. Lightly drag the cotton swab across the agar surface in a zigzag pattern. Dispose of the swab as shown in Figure 1-9c. Incubate the plate in an inverted position for the assigned time at the appropriate temperature. Be sure to label the base with your name, date and sample.

FIGURE 1-9b   **Streak Plate Inoculation**
This inoculation pattern is performed for isolation of two or more bacterial species in a mixed culture with suspected high cell density. Hold the swab comfortably in one hand and lift the lid of the Petri dish with the other. Use the lid as a shield to protect the agar from aerial contamination. Lightly drag the cotton swab back and forth across the agar surface in one quadrant of the plate. Further streaking is performed with a loop as described in Figure 4-1, page 53. Dispose of the swab as shown in Figure 1-9c. Incubate the plate in an inverted position for the assigned time at the appropriate temperature. Be sure to label the base with your name, date and sample.

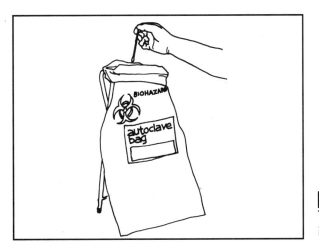

FIGURE 1-9c   **Dispose of the Swab**
The swab is contaminated and must be disposed of properly in a Biohazard container destined for autoclaving.

SPOT INOCULATION OF AGAR PLATES

Sometimes, an agar plate may be used to grow several different specimens at once. This is a typical practice with plated *differential* media (*i.e.*, media designed to differentiate organisms based on growth characteristics). Prior to beginning the transfer, the plate may be divided into as many as four sectors using a marking pen (some plates already have marks on the base for this purpose). Each may then be inoculated with a different organism. Inoculation involves touching the loop to the agar surface once so that growth is restricted to a single spot — hence the name "spot inoculation."

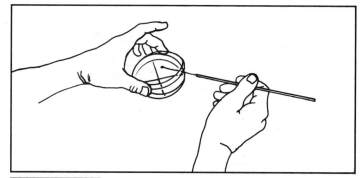

FIGURE 1-10a   **Inoculate the Medium**

Lift the lid of the sterile agar plate and use it as a shield to prevent aerial contamination. Then touch the agar surface in the center of the sector. Remove the loop and replace the lid.

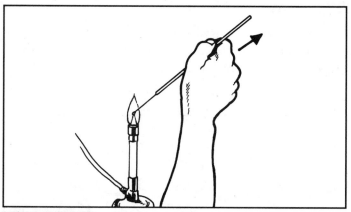

FIGURE 1-10b   **Flame the Loop**

Sterilize your loop as before. It is especially important to flame it from base to tip now because the loop has lots of bacteria on it. Label the plate's base with your name, date and organism(s) inoculated. Incubate the plate in an inverted position for the assigned time at the appropriate temperature.

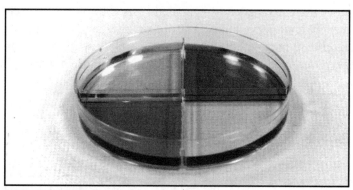

FIGURE 1-10c   **Multiple Tests May Be Run**

Petri plates may be purchased which have built-in partitions, as in this photo. Four different organisms may be spot inoculated if all wells contain the same medium. Or, as shown here, four different tests can be run simultaneously on the same organism if each well contains a different medium.

INOCULATION OF AGAR PLATES WITH A PIPETTE — THE SPREAD PLATE TECHNIQUE

The spread plate technique is often used with quantitative procedures, but may also be used for isolation of a particular organism from a mixed culture.

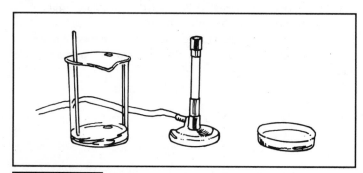

FIGURE 1-11a   **Safety First**

The spread plate technique requires a Bunsen burner, a beaker with alcohol, a glass spreading rod and the plate. Position these components in your work area as shown: ethyl alcohol, flame, and plate. This arrangement reduces the chance of accidentally catching the alcohol on fire.

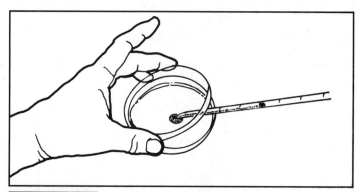

FIGURE 1-11b   **Inoculate the Plate**

Lift the plate's lid and use it as a shield to protect from airborne contamination. Insert the pipette and dispense the correct volume (often 0.1 mL) onto the center of the agar surface. From this point, the remainder of steps should be completed within about 15 seconds to prevent the inoculum from soaking into the agar.

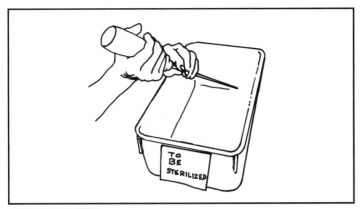

FIGURE 1-11c   **Dispose of the Pipette**

The pipette is contaminated with microbes and must be correctly disposed of. Each lab has its own specific procedures and your instructor will advise you what to do. Shown here is a glass pipette being placed in pipette disposal container containing a small amount of amphyl disinfectant. Disposable pipettes must be placed in an appropriate biohazard container. In either case, be careful when removing the pipette from the mechanical pipettor. There is danger of culture dripping from the pipette or of breaking the glass.

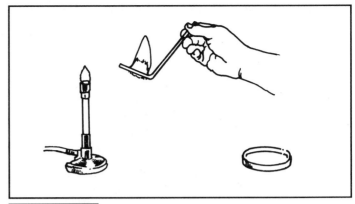

FIGURE 1-11d   **Sterilize the Glass Rod**

Remove the glass spreading rod from the alcohol and pass it through the flame to ignite the alcohol. Remove the rod from the flame and allow the alcohol to burn off completely. Do not leave the rod in the flame; the combination of the alcohol and brief flaming are sufficient to sterilize it. *Be careful not to drop any flaming alcohol on the work surface. Be especially careful not to drop flaming alcohol back into the alcohol container!*

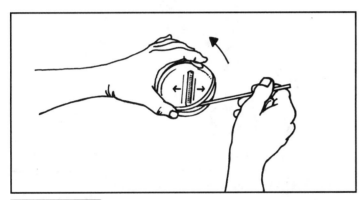

FIGURE 1-11e   **Spread the Inoculum**

After the flame has gone out on the rod, lift the lid of the plate and use it as a shield from airborne contamination. Then, touch the rod to the agar surface away from the inoculum in order to cool it. To spread the inoculum, hold the plate lid with the base of your thumb and index finger, and use the tip of your thumb and middle finger to rotate the base. At the same time, move the rod in a back-and-forth motion across the agar surface. After a couple of turns, do one last turn with the rod next to the plate's edge.

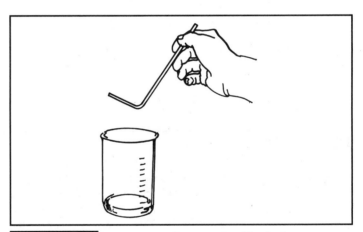

FIGURE 1-11f   **Return the Rod to the Alcohol**

Remove the rod from the plate and replace the lid. Return the rod to the alcohol in preparation for the next inoculation. Label the plate base with your name, date, organism, and any other relevant information. Incubate the plate in an inverted position at the appropriate temperature for the assigned time. (If you plated a large volume of inoculum, wait a few minutes and allow it to soak in before inverting the plate.)

STAB INOCULATION OF AGAR TUBES USING AN INOCULATING NEEDLE

Stab inoculations of agar tubes are used for several types of differential media (usually to examine growth under anaerobic conditions). A stab is *not* used to produce a culture of microbes for transfer to another medium.

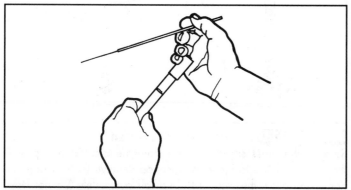

FIGURE 1-12a   Hold the Cap in Your Pinkie Finger
While keeping the needle hand still, bring the agar tube towards it. Use the pinkie finger of your needle hand to remove and hold its cap. (The cap should be loosened prior to the transfer, especially if it's a screw top cap.)

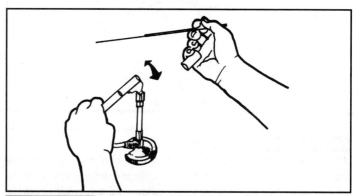

FIGURE 1-12b   Sterilize the Tube
Pass the lip of the tube quickly through the flame two or three times to sterilize the glass and the surrounding air. The tube should be held on an angle to prevent contamination from above. Keep your needle hand still.

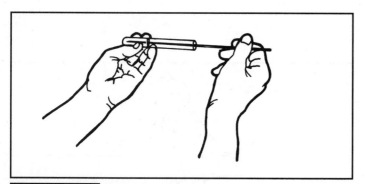

FIGURE 1-12c   Stab the Agar
Hold the open tube at an angle to prevent aerial contamination. Carefully move the agar tube over the needle wire. Insert the needle into the agar, then withdraw the tube carefully (ideally) along the same stab line. Be especially careful when removing the tube not to catch the needle tip on the tube lip. This springing action of the needle creates bacterial aerosols.

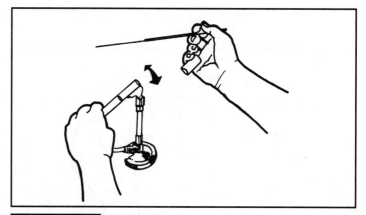

FIGURE 1-12d   Sterilize the Tube Again
Flame the tube lip as before. Keep your needle hand still.

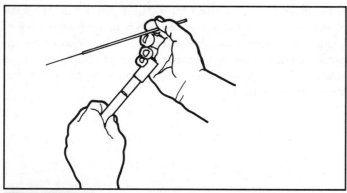

FIGURE 1-12e   Replace the Cap
Keeping the needle hand still (remember, it has growth on it), move the tube to replace its cap. The cap at this point doesn't need to be on firmly — just enough to cover the tube.

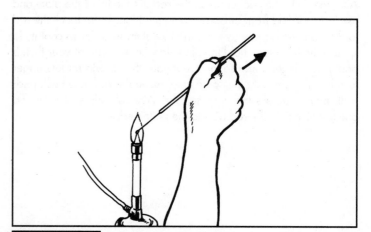

FIGURE 1-12f   Flame the Needle
Sterilize the needle as before by incinerating it in the Bunsen burner flame. Label the tube with your name, date and organism. Incubate at the appropriate temperature for the assigned time.

FISHTAIL INOCULATION OF AGAR SLANTS

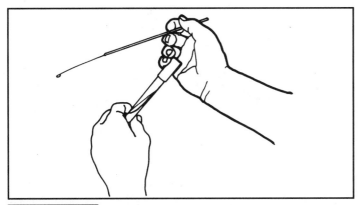

FIGURE 1-13a   **Hold the Cap in Your Pinkie Finger**
While keeping the loop hand still, bring the agar slant tube towards it. Use your pinkie finger to remove and hold its cap. (The cap should be loosened prior to the transfer, especially if it's a screw top cap.)

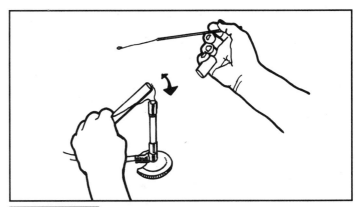

FIGURE 1-13b   **Sterilize the Tube**
Pass the lip of the tube quickly through the flame two or three times to sterilize the glass and the surrounding air. The tube should be held on an angle to prevent contamination from above. Keep your loop hand still.

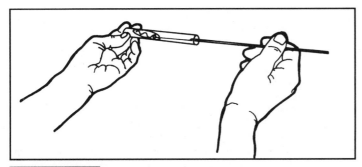

FIGURE 1-13c   **Inoculate the Agar**
With the agar surface facing upward, hold the open tube at an angle to prevent aerial contamination. Carefully move the agar tube over the loop. Beginning at the bottom of the exposed agar surface, drag the loop in a zigzag pattern as the tube is removed. Be especially careful when removing the tube not to catch the loop tip on the tube lip. This springing action of the loop creates bacterial aerosols.

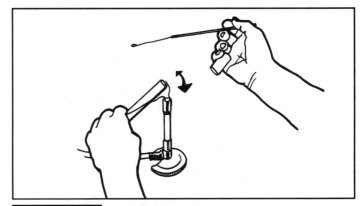

FIGURE 1-13d   **Sterilize the Tube Again**
Flame the tube lip as before. Keep your loop hand still.

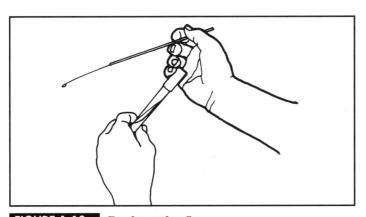

FIGURE 1-13e   **Replace the Cap**
Keeping the loop hand still (remember, it has growth on it), move the tube to replace its cap. The cap at this point doesn't need to be on firmly — just enough to cover the tube.

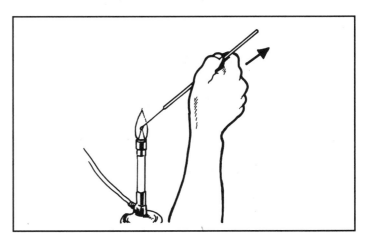

FIGURE 1-13f   **Flame the Loop**
Sterilize the loop as before by incinerating it in the Bunsen burner flame. Label the tube with your name, date and organism. Incubate at the appropriate temperature for the assigned time.

INOCULATION OF BROTH TUBES WITH A LOOP OR NEEDLE

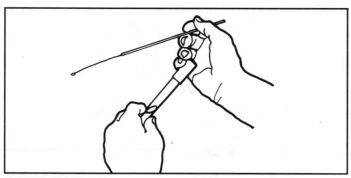

FIGURE 1-14a   Hold the Cap in Your Pinkie Finger
[Note: since loops and needles are handled in the same way, we refer only to a loop in the following instructions for ease of reading.] While keeping the loop hand still, bring the broth tube towards it. Use your pinkie finger to remove and hold its cap. (The cap should be loosened prior to the transfer, especially if it's a screw top cap.)

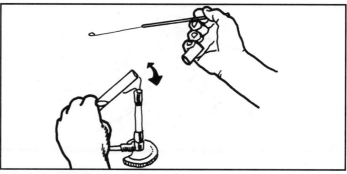

FIGURE 1-14b   Sterilize the Tube
Pass the lip of the tube quickly through the flame two or three times to sterilize the glass and the surrounding air. The tube should be held on an angle to prevent contamination from above. Keep your loop hand still.

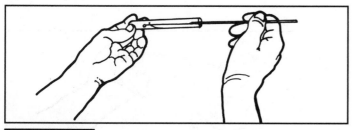

FIGURE 1-14c   Inoculate the Broth
Hold the open tube at an angle to prevent aerial contamination. Carefully move the broth tube over the wire. Gently swirl the loop to dislodge microbes.

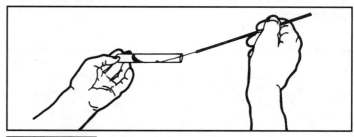

FIGURE 1-14d   Remove Excess Broth
Withdraw the tube from over the loop. Before completely removing it, touch the loop tip to the glass to remove any excess broth. Be especially careful when removing the tube not to catch the loop tip on the tube lip. This springing action of the wire creates bacterial aerosols.

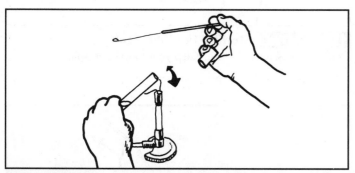

FIGURE 1-14e   Sterilize the Tube Again
Flame the tube lip as before. Keep your loop hand still.

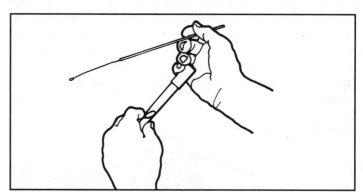

FIGURE 1-14f   Replace the Cap
Keeping the loop hand still (remember, it has growth on it), move the tube to replace its cap. The cap at this point doesn't need to be on firmly — just enough to cover the tube.

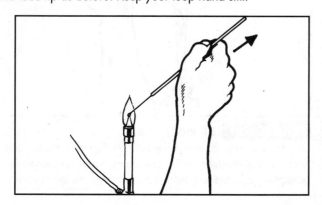

FIGURE 1-14g   Flame the Loop/Needle
Sterilize the loop as before by incinerating it in the Bunsen burner flame. Label the tube with your name, date and organism. Incubate at the appropriate temperature for the assigned time.

INOCULATION OF BROTH TUBES WITH A PIPETTE

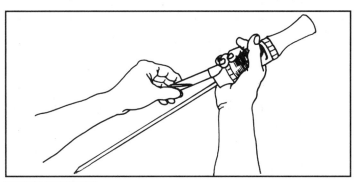

FIGURE 1-15a **Hold the Cap in Your Pinkie Finger**
While keeping the pipette hand still, bring the broth tube towards it. Use your pinkie finger to remove and hold its cap. (The cap should be loosened prior to the transfer, especially if it's a screw top cap.)

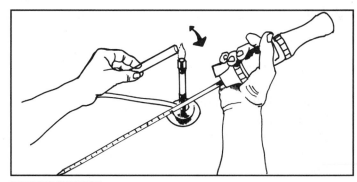

FIGURE 1-15b **Sterilize the Tube**
Pass the lip of the tube quickly through the flame two or three times to sterilize the glass and the surrounding air. The tube should be held on an angle to prevent contamination from above. Keep your pipette hand still.

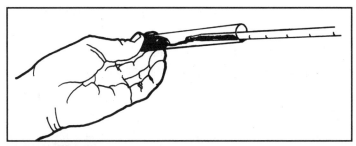

FIGURE 1-15c **Inoculate the Broth**
Hold the open tube at an angle to prevent aerial contamination. Insert the pipette tip and dispense the correct volume of inoculum.

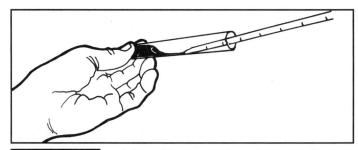

FIGURE 1-15d **Remove Excess Broth**
Withdraw the tube from over the pipette. Before completely removing it, touch the pipette tip to the glass to remove the any excess broth. Completely remove the pipette, but avoid waving it around. This can create aerosols.

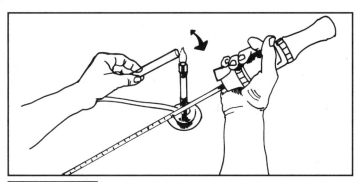

FIGURE 1-15e **Sterilize the Tube Again**
Flame the tube lip as before. Keep your pipette hand still.

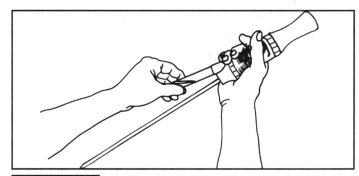

FIGURE 1-15f **Replace the Cap**
Keeping the pipette hand still, move the tube to replace its cap. The cap at this point doesn't need to be on firmly — just enough to cover the tube.

FIGURE 1-15g **Dispose of the Pipette**
The pipette is contaminated with microbes and must be correctly disposed of. Each lab has its own specific procedures and your instructor will advise you what to do. Shown here is a glass pipette being placed in pipette disposal container containing a small amount of amphyl disinfectant. Disposable pipettes must be placed in an appropriate biohazard container. In either case, be careful when removing the pipette from the mechanical pipettor. There is danger of culture dripping from the pipette or of breaking the glass.

Aseptic Transfers

MATERIALS

Inoculating loop
Sterile cotton swab
Four nutrient broth tubes
Five nutrient agar slants
One nutrient agar plate
Marking pen and labels
Vortex mixer
Slant cultures of:
   Bacillus subtilis
   Escherichia coli
   Micrococcus luteus
Broth culture of:
   Staphylococcus epidermidis
Plate culture of:
   Micrococcus roseus

PROTOCOL

Day One

1. While referring to "Microbial Transfer Methods" above, make the following transfers:

    a. *B. subtilis* slant to slant and broth using a loop (Figs. 1-4, 1-13 and 1-14).

    b. *E. coli* slant to slant and broth using a loop.

    c. *M. luteus* slant to slant and broth using a loop.

    d. *S. epidermidis* broth to slant using an inoculating loop (Figs 1-5 and 1-13).

    e. *S. epidermidis* broth to plate using a sterile swab. (Make this a simple zigzag pattern across the plate as in Figure 1-9a.) Dispose of the swab in an autoclave container.

    f. *M. roseus* plate to slant and broth. (For each transfer choose a well isolated colony and just touch the center with the loop as in Figures 1-3, 1-13 and 1-14.)

2. Label all tubes and plates clearly with your name, the organisms' names, and the date. Tape the lid of the plate down to prevent it from accidentally opening.

3. Incubate *M. luteus* and *M. roseus* at 25°C and the rest at 37°C until next class. Be sure to invert the plate for incubation.

Day Two

1. Remove the media from the incubator and examine the growth.

2. Dispose of all materials in the appropriate autoclave containers.

PRECAUTIONS

⚠ Be organized. Arrange all media in advance and clearly label it. Be sure not to place the labels in such a way as to obscure your view of the inside of the tube or plate.

⚠ Allow time for the loop to cool before touching the bacterial growth.

⚠ Take your time. Work efficiently, but *do not hurry*. You are handling potentially dangerous bacteria.

REFERENCES

Barkley, W. Emmett and John H. Richardson. 1994. Chapter 29 in *Methods for General and Molecular Bacteriology*. American Society for Microbiology, Washington, D.C.

Boyer, Rodney F. 1993. *Modern Experimental Biochemistry, 2nd Ed.* Benjamin/Cummings Publishing Company, Inc., Redwood City, CA.

Claus, G. William. 1989. Chapter 2 *in Understanding Microbes – A Laboratory Textbook for Microbiology*. W.H. Freeman and Company, New York, NY.

Darlow, H. M. 1969. Chapter VI in *Methods in Microbiology, Volume 1*. Edited by J. R. Norris and D. W. Ribbins. Academic Press, Ltd., London.

Fleming, Diane O. 1995. Chapter 13 in *Laboratory Safety – Principles and Practices, 2nd Ed.* Edited by Diane O. Fleming, John H. Richardson, Jerry J. Tulis and Donald Vesley. American Society for Microbiology, Washington, D.C.

Koneman, Elmer W., Stephen D. Allen, William M. Janda, Paul C. Schreckenberger and Washington C. Winn, Jr. 1997. Chapter 2 in *Color Atlas and Textbook of Diagnostic Microbiology, 5th Ed.* Lippincott-Raven Publishers, Philadelphia, PA.

Murray, Patrick R., Ellen Jo Baron, Michael A. Pfaller, Fred C. Tenover, and Robert H. Yolken. 1995. *Manual of Clinical Microbiology, 6th Ed.* American Society for Microbiology, Washington, D.C.

Power, David A. and Peggy J. McCuen. 1988. *Manual of BBL® Products and Laboratory Procedures, 6th Ed.* Becton Dickinson Microbiology Systems, Cockeysville, MD.

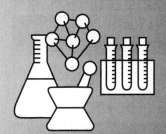

Cultivation and Growth of Bacterial Cultures

*I*dentification of a microbe begins with determining some basic characteristics of the organism: gross appearance of its growth, environment (including nutrients) necessary to support its growth, and staining properties. In this section you will be learning about the various growth characteristics and environments; staining is covered in Section Three.

GROSS APPEARANCE OF BACTERIAL GROWTH

A microbiologist's first encounter with a particular microbe is typically as a colony on an agar plate. In fact, the first way of distinguishing one bacterial species from another is by recognizing differences in the appearance of their growth. Growth patterns in broth culture may also be helpful in determining characteristics of the organism.

Distinguishing different growth patterns is a skill that you learn in this section and will employ as you progress through the semester, especially in the determination of your unknown species (Section Six). Note the growth characteristics of all the organisms you see this semester; you will be glad you did!

Exercise 2–1	**Bacterial Colony Morphology**

MATERIALS

First Lab Period

Six nutrient agar plates
A sterile cotton swab in sterile saline

Second Lab Period

Recommended:
Streak plate culture of *Chromobacterium violaceum*
Streak plate culture of *Micrococcus roseus*
Streak plate culture of *Pseudomonas aeruginosa*
Streak plate culture of *Serratia marcescens*
A colony counter
A six inch plastic ruler to measure colony diameters

PROCEDURE

Day One

1. Inoculate the five plates as follows:

   a. Open one plate and expose it to the air for 30 minutes (or longer, if convenient).

   b. Use the cotton swab to sample your desk area, then streak the second agar plate as in Figure 1-9a.

   c. Cough several times on the agar surface of the third plate.

   d. Touch the agar surface of the fourth agar plate lightly with your fingertips. It is best if you *haven't* washed your hands recently.

Photographic Atlas Reference Page 1

ORGANISM / PLATE	COLONY DESCRIPTION

  e. Remove the lid of the fifth agar plate and vigorously scratch your head over it.

  f. Do not inoculate the sixth plate. It is a control to ensure your agar plates are sterile.

2. Label the base of each plate with your name, the date and the type of exposure it has received.

3. Incubate the six plates in an inverted position at 37°C for at least one day.

Day Two

1. Figure 2-1 shows some typical colonial characteristics and their descriptive terms. Use these terms, and others by your instructor, to describe the colonies on your plates from the first lab as well as those supplied. Measure colony diameters with the ruler and include them with your descriptions in the table below.

2. It may be helpful to use a colony counter (Figure 2-2) to assist in viewing the plates. Colony counters are equipped with a magnifying glass and light to allow observation of detail and optical properties of the growth.

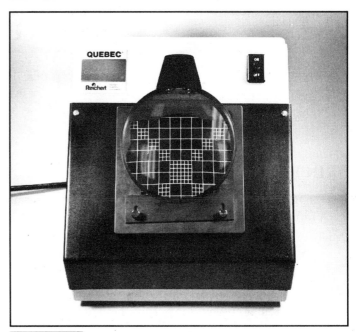

FIGURE 2-2   **The Colony Counter**
The colony counter may be useful in seeing subtle differences between similar colonies. The grid in the background is composed of 1 cm squares.

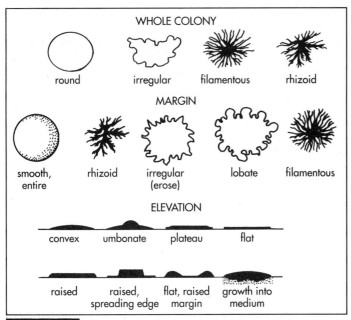

FIGURE 2-1   **A Sampling of Bacterial Colony Features**
Use these terms to describe colonial morphology. Color, surface characteristics (dull or shiny), consistency (dry, butyrous-buttery, or moist) and optical properties (opaque or translucent), should also be included in colony morphology descriptions.

PRECAUTIONS

⚠ Organisms that you find on your plates in this exercise are common laboratory contaminants, so observe them carefully.

⚠ Don't expect to see examples of all the descriptive terms in this laboratory exercise.

⚠ Save your plates for the "Growth Patterns in Broth" exercise.

REFERENCES

Claus, G. William. 1989. Chapter 14 in *Understanding Microbes – A Laboratory Textbook for Microbiology*. W.H. Freeman and Company, New York, NY.

Collins, C.H., Patricia M. Lyne and J.M. Grange. 1995. Chapter 6 in *Collins and Lyne's Microbiological Methods, 7th Ed.* Butterworth-Heineman, Oxford.

Koneman, Elmer W., Stephen D. Allen, William M. Janda, Paul C. Schreckenberger and Washington C. Winn, Jr. 1997. Chapter 2 in *Color Atlas and Textbook of Diagnostic Microbiology, 5th Ed.* J.B. Lippincott Company, Philadelphia, PA.

Exercise 2–2

Growth Patterns in Broth

MATERIALS

Five nutrient broth tubes
Your plates from the Bacterial Colony Morphology exercise

PROCEDURE

1. Using your loop and the plates from the Bacterial Colony Morphology exercise, aseptically transfer growth into five different nutrient broth tubes. (See Figs. 1-3 and 1-14 for the methods.)

2. Label each tube with your name, the date, the medium and a description of the colony (*e.g.,* small, white circular colony) for purposes of identification.

3. Incubate the tubes at 37°C for at least one day.

4. After incubation, examine the tubes and describe differences in growth. Record the results in the table below. Page 5 in the *Photographic Atlas* lists some of the descriptive terms you can use.

PRECAUTIONS

⚠ Check with your classmates to see if they got growth patterns consistent with those of your organisms.

⚠ You may not see examples of all growth patterns from your samples. Check to see if your classmates got any different ones.

⚠ You may see more than one pattern in a single tube, as with *Chromobacterium violaceum* in the *Photographic Atlas* which illustrates a ring, sediment and fine turbidity.

Photographic Atlas Reference Page 5

REFERENCE

Claus, G. William. 1989. Chapter 17 in *Understanding Microbes – A Laboratory Textbook for Microbiology.* W.H. Freeman and Company, New York, NY.

ORGANISM	DESCRIPTION OF GROWTH IN BROTH

AEROTOLERANCE

Aerotolerance is the ability of a microbe to grow in the presence of oxygen. This is a fundamental characteristic of bacteria that may be used in identification, but also must be considered before the bacteria can be cultivated. It is so basic, in fact, that *Bergey's Manual of Determinative Bacteriology, 9th Ed.* and *Bergey's Manual of Systematic Bacteriology* include aerotolerance category in many of their section titles. Aerotolerance categories include *obligate aerobes, facultative anaerobes, aerotolerant anaerobes, microaerophiles,* and *obligate anaerobes.* This exercise introduces you to a sampling of these groups and to a means of determining them.

Exercise 2-3   Agar Deep Stabs

MATERIALS

Four nutrient agar deep tubes
Recommended organisms:
   Clostridium spp. in thioglycolate broth
   Moraxella spp. in nutrient broth
   Providencia spp. in nutrient broth

TEST PROTOCOL

1. Stab inoculate three tubes with different organisms (Figs. 1-5 and 1-12). Use a heavy inoculum.

2. Stab the fourth tube with your sterile needle to act as a control.

3. Label each tube with your name, the date, and the medium.

4. Incubate the tubes at 37°C for at least one day.

5. After incubation, examine the tubes and try to determine the aerotolerance category of each. Record the results in the table below.

PRECAUTIONS

Photographic Atlas Reference Page 6

⚠ Use your control tube to help you discriminate between growth along the stab line and the stab line itself.

⚠ Try to enter and exit along the same stab line.

REFERENCES

Baron, Ellen Jo, Lance R. Peterson, and Sydney M. Finegold. 1994. Chapter 9 in *Bailey and Scott's Diagnostic Microbiology, 9th Ed.* Mosby-Year Book, Inc. St. Louis, MO 63146.

Koneman, Elmer W., Stephen D. Allen, William M. Janda, Paul C. Schreckenberger and Washington C. Winn, Jr. 1997. Chapter 14 in *Color Atlas and Textbook of Diagnostic Microbiology, 5th Ed.* J.B. Lippincott Company, Philadelphia, PA.

ORGANISM	DESCRIPTION OF GROWTH	AEROTOLERANCE CATEGORY

ANAEROBIC CULTURE METHODS

Growth of aerobic organisms requires little beyond supplying the appropriate nutrients for survival. However, cultivation of anaerobes presents an extra challenge. In addition to supplying necessary nutrients, growing strict anaerobes requires the creation of an environment devoid of oxygen. This can be accomplished in many different ways, depending on the laboratory's resources and the demands of the particular work being done. Two methods are considered here: thioglycolate broth and the anaerobic jar. Thioglycolate broth is a special medium which creates an anaerobic environment *within the tube* and is incubated

aerobically. The anaerobic jar creates an artificial *anaerobic environment* in which ordinary media can be incubated.

Although these methods are primarily for cultivation of anaerobes, specific aerotolerance categories may also be determined. In these exercises you will be using both the anaerobic jar and thioglycolate broth to compare and categorize four organisms based on their oxygen tolerance properties.

Exercise 2–4

Thioglycolate Broth

MATERIALS

Five thioglycolate tubes
Recommended organisms:
  Pseudomonas aeruginosa
  Clostridium sporogenes
  Staphylococcus aureus
  Lactobacillus plantarum

TEST PROTOCOL

1. Inoculate four thioglycolate tubes with the test organisms. Use the fifth tube as an uninoculated control.

2. Label each tube with your name, the date, medium and organism.

3. Incubate the tubes at 37°C for at least one day.

4. After incubation, examine the tubes and determine the oxygen tolerance category of each. Record the results in the table below.

PRECAUTIONS

⚠ Thioglycolate tubes should have a pink band at the surface (the aerobic zone). If none is visible, the medium is of suspect quality.

⚠ If, after storage, more than 30% of the medium is pink, the tubes should be boiled with caps loosened to remove the excess oxygen. Tighten the caps and cool to room temperature before use.

> *Photographic Atlas Reference Page 7*

REFERENCES

Baron, Ellen Jo, Lance R. Peterson, and Sydney M. Finegold. 1994. Chapter 9 in *Bailey and Scott's Diagnostic Microbiology, 9th Ed.* Mosby-Year Book, Inc. St. Louis, MO 63146.

DIFCO Laboratories. 1984. Page 951 in *DIFCO Manual, 10th Ed.* DIFCO Laboratories, Detroit, MI.

Power, David A. and Peggy J. McCuen. 1988. Page 261 in *Manual of BBL® Products and Laboratory Procedures, 6th Ed.* Becton Dickinson Microbiology Systems, Cockeysville, MD.

ORGANISM	LOCATION OF GROWTH IN BROTH	OXYGEN TOLERANCE CATEGORY

Exercise 2-5

Anaerobic Jar

MATERIALS

Two nutrient agar plates
One anaerobic jar with gas generator packet (Becton Dickinson Microbiology Systems, Sparks, MD)
Recommended organisms:
Pseudomonas aeruginosa
Clostridium sporogenes
Staphylococcus aureus
Lactobacillus plantarum

TEST PROTOCOL

1. Divide each nutrient agar plate into four sectors with your marking pen.

2. Use your loop to spot inoculate (Fig 1-10) the four organisms into different sectors of each plate.

3. Label each plate with your name, the organisms' names in each sector, the date, medium and incubation conditions.

4. Place *one* plate in the anaerobic jar in an inverted position. (Other students will do the same with their plates.)

5. When the jar is filled, discharge the packet as follows (or follow the instructions on your packet).
   a. Stick the methylene blue strip on the wall of the jar.
   b. Open the packet and dispense 10 mL of distilled water.
   c. Place the open packet with the writing facing inward into the jar.
   d. Immediately close the jar.

6. Incubate the anaerobic jar at 37°C for at least one day. Incubate the other plate aerobically at the same temperature for the same length of time.

7. After incubation, examine and compare the plates. Record the results in the table below.

Photographic Atlas Reference Page 8

PRECAUTIONS

⚠ Make certain the gas generator packet for the anaerobic jar has not expired.

⚠ Once water is added to the gas generator packet, close the lid immediately.

⚠ If no condensation is seen inside the jar within 30 minutes, check the seal on the jar lid and repeat the process.

⚠ After incubation, check the methylene blue strip before opening the jar. It should be white. If it is not, conditions may not have been anaerobic.

REFERENCES

Allen, Stephen D., Jean A. Siders, and Linda M. Marler. 1985. Chapter 37 in *Manual of Clinical Microbiology, 4th Ed.* Edited by Edwin H. Lennette, Albert Balows, William J. Hausler, Jr., and H. Jean Shadomy. American Society for Microbiology, Washington, D.C.

Baron, Ellen Jo, Lance R. Peterson, and Sydney M. Finegold. 1994. Chapter 9 in *Bailey and Scott's Diagnostic Microbiology, 9th Ed.* Mosby-Year Book, Inc. St. Louis, MO 63146.

Koneman, Elmer W., Stephen D. Allen, William M. Janda, Paul C. Schreckenberger and Washington C. Winn, Jr. 1997. Chapter 14 in *Color Atlas and Textbook of Diagnostic Microbiology, 5th Ed.* J.B. Lippincott Company, Philadelphia, PA.

Power, David A. and Peggy J. McCuen. 1988. Page 311 in *Manual of BBL® Products and Laboratory Procedures, 6th Ed.* Becton Dickinson Microbiology Systems, Cockeysville, MD.

ORGANISM	AEROBIC GROWTH (+ or −)	ANAEROBIC GROWTH (+ or −)	OXYGEN TOLERANCE CATEGORY

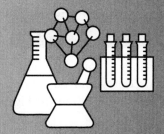

Microscopy and Staining

*I*t probably could go without saying — but we'll say it anyway — that microbiology as a biological discipline would not be what it is today without microscopes or stains. Our ability to visualize, sometimes in great detail, the form and structure of microbes too small or transparent to otherwise be seen is due almost exclusively to developments in microscopy and staining techniques. In this section, you will learn (or refine) your microscope skills. Then, you will learn simple and more sophisticated bacterial staining techniques.

BASIC LIGHT MICROSCOPY

Proper use of the microscope is absolutely essential for your success in microbiology. Fortunately, with practice and by following a few simple guidelines, you can achieve satisfactory results quickly. Since student labs may be supplied with a variety of microscopes, your instructor may supplement the following procedures and guidelines with instructions specific to your equipment. Refer to Figure 3-1 as you read the following.

Guidelines for Use of the Light Microscope

1. Carry your microscope to your work station using both hands — one hand grasping the microscope's arm, the other supporting the microscope beneath its base.

2. Use cotton swabs to clean the objective, ocular, and condenser lenses. Lens paper is useful for gently blotting the oil from the oil immersion lens, but cotton swabs dipped in pure ethanol or commercial lens cleaning solution should be used for final cleaning of any objective. To clean an ocular, moisten the cotton with appropriate liquid and gently wipe in a spiral motion starting at the center of the lens and working outward. Always wipe lenses dry with a clean cotton swab.

3. Raise the condenser to its maximum position nearly even with the stage and open the iris diaphragm.

4. Plug in the microscope and turn the lamp on. Adjust the light intensity slowly to its maximum.

5. Move the low power objective (usually 4X) into position.

6. Place the slide on the stage in the mechanical slide holder. Center the specimen over the opening in the stage.

7. If using a binocular microscope, adjust the position of the two oculars to match your own interpupillary distance.

8. Use the coarse focus adjustment knob to bring the image into focus. Bring the image into sharpest focus using the fine focus adjustment knob. Then, observe the specimen with your eyes relaxed and slightly above the oculars to allow the images to fuse into one.

9. If you are using a binocular microscope, you may adjust the oculars' focus to compensate for differences in visual acuity of your two eyes.

10. Adjust the iris diaphragm and condenser position to produce optimum illumination, contrast and image.

11. Scan the specimen to locate a promising region to examine in more detail.

12. If you are observing a nonbacterial specimen, progress through the objectives until you see the degree of structural detail necessary for your purposes. You will need to adjust the fine focus and illumination for each objective. Before advancing to the next objective, be sure to position a desirable portion of the specimen in the center of the field or you risk "losing" it at the higher magnification.

13. If you are working with a bacterial smear, you will need to use the oil immersion lens. Work through the medium (10X), then high dry (40X) objectives, adjusting the fine focus and illumination for each. Before advancing to the next objective, be sure to position a desirable portion of the specimen in the center of the field or you risk "losing" it at the higher magnification.

14. To use the oil immersion lens, rotate the nosepiece to a position midway between the high dry and oil immersion lens. Then, place a drop of immersion oil on the specimen. *Be careful not to get any oil on the microscope or its lenses, and be sure to clean it up if you do.* Rotate the oil lens so its tip is submerged in the oil drop. Focus and adjust the illumination to maximize image quality. (Note: Do not move the stage down at this point or you will lose your focal plane. On a properly adjusted microscope the oil lens and the high dry lens have the same focal plane. Therefore, when a specimen is in focus on high dry, the oil lens, although longer, will not touch the slide when rotated into position.

15. When finished, lower the stage (or raise the objective) and remove the slide. Dispose of the freshly-prepared slides in a jar of disinfectant; return permanent slides to storage after wiping off any oil.

16. When finished for the day, be sure to do the following:

    a. Center the mechanical stage.

    b. Lower the light intensity to its minimum, then turn off the light.

    c. Wrap the electrical cord according to your particular lab rules.

d. Clean any oil off the lenses, stage, *etc*. Be sure to use only cotton swabs or lens paper for cleaning any of the optical surfaces of the microscope (see #2).

e. Return the microscope to its appropriate storage place.

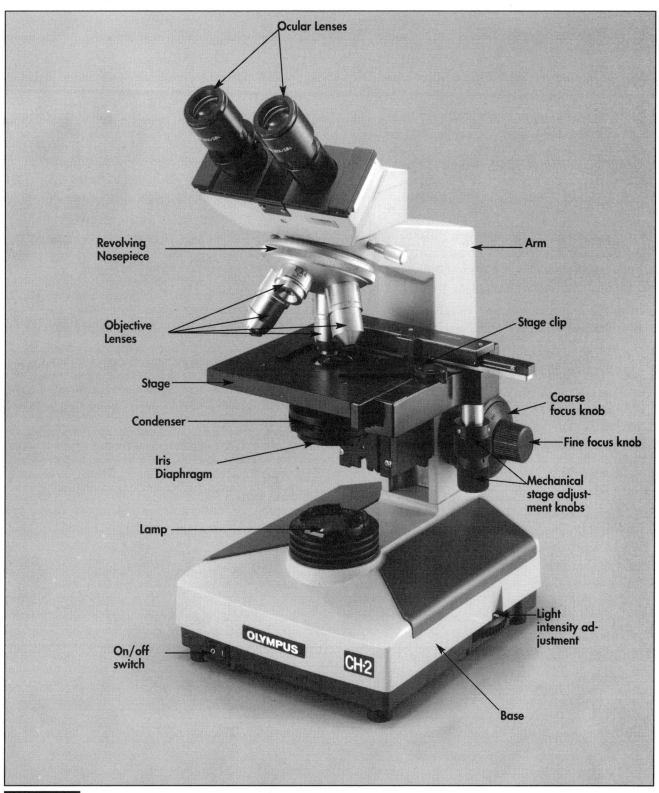

FIGURE 3-1   **A Binocular Compound Microscope**

A quality microscope is essential to the field of microbiology.   (Photograph courtesy of Olympus America, Inc.)

Exercise 3-1

Examination of Prepared Microscope Slides

MATERIALS

Prepared slides of a variety of protozoans and fungi
Microscope
Immersion oil
Cotton swabs
Lens paper

INSTRUCTIONS

1. Obtain a microscope and place it on the table or workspace. Check to be sure the stage is all the way down and the low power objective is in place.

2. Place a prepared slide on the microscope stage and center the specimen under the low power objective.

3. Following the instructions given in "Guidelines for Use of the Light Microscope" above, bring the specimen into focus at the highest magnification which allows you to see the entire structure you want to view.

4. Practice scanning with the mechanical stage until you are satisfied that you have seen everything interesting to see.

5. Repeat with as many slides as you have time for.

6. When you are finished, blot the oil from the oil immersion lens with a lens paper and do a final cleaning with a cotton swab and alcohol or lens cleaning solution. Dry the lenses with a clean cotton swab.

7. Return all lenses and adjustments to their storage positions before putting the microscope away.

PRECAUTIONS

⚠ Avoid using too much illumination, or you will lose contrast.

⚠ If you do "lose" your specimen, it is generally faster to go to low power and work your way back to where you were than to search aimlessly at the higher magnification.

⚠ Use caution when focusing with the high dry and oil lenses to avoid driving the lens through the microscope slide. Use only the fine focus knob.

⚠ Oil bubbles or the oil immersion lens tip being above the oil produces a poor quality image. Be certain it is *immersed* in oil.

⚠ If you are unable to get the specimen in focus using the oil immersion lens, you may have the slide in upside down.

REFERENCES

Baron, Ellen Jo, Lance R. Peterson, and Sydney M. Finegold. 1994. Chapter 8 in *Bailey and Scott's Diagnostic Microbiology, 9th Ed.* Mosby-Year Book, Inc. St. Louis, MO 63146.

Chapin, Kimberle. 1995. Chapter 4 in *Manual of Clinical Microbiology, 6th Ed.*, edited by Patrick R. Murray, Ellen Jo Baron, Michael A. Pfaller, Fred C. Tenover, and Robert H. Yolken. American Society for Microbiology, Washington, D.C.

Murray, R.G.E. and Carl F. Robinow. 1994. Chapter 1 in *Methods for General and Molecular Bacteriology*, edited by Philipp Gerhardt, R.G.E. Murray, Willis A. Wood, and Noel R. Krieg. American Society for Microbiology, Washington, D.C.

Calibration of the Ocular Micrometer

INTRODUCTION AND SAMPLE CALIBRATIONS

The dimensions of a microbial cell can be determined with the microscope using a device called an *ocular micrometer* (Fig. 3-2). An ocular micrometer is a type of ruler, installed in the eyepiece, composed of uniformly spaced graduations of unknown size. To be of use, therefore, the ocular micrometer must be calibrated. A *stage micrometer* (a slide with a microscopic ruler etched into it) is the device used for this purpose (Fig. 3-3). Because changing from one objective lens to another changes the magnification of the image (relative to the ocular micrometer which remains constant) calibration must be performed for each magnification. Calibrations for each objective lens are valid as long as the ocular and objective lenses are not changed on the microscope.

The stage micrometer used in Figure 3-3 is 2.2 mm long, with major increments of 0.1 mm (100 μm), two of which are divided into 0.01 mm (10 μm) increments. To calibrate the ocular micrometer, place the stage micrometer on the microscope so that it is in focus on low power and superimposed by the ocular micrometer (Fig. 3-4). Align the lefthand line of the ocular micrometer with the first mark on the stage micrometer. Without moving anything but your eyes, look to the right and find other lines that correspond exactly. In Figure 3-4, note that the first

mark on the stage micrometer that lines up with an ocular micrometer line is at 0.8 mm. Note also the 25 ocular micrometer spaces (ocular units) within that 0.8 mm span. You will also notice that 1.5 mm is equivalent to 47 ocular units. These values have been entered for you in the table below.

STAGE MICROMETER	OCULAR MICROMETER
0.8 mm	25 ocular units
1.5 mm	47 ocular units

To determine the value for each ocular unit, divide the distance (from the stage micrometer) by the number of ocular units (from the ocular micrometer) to which it corresponds. Multiply by 1000μm/mm to convert the units to μm.

$$\frac{0.8 \text{ mm}}{25 \text{ ocular units}} = 0.032 \text{ mm/OU} = 32\mu m/OU$$

$$\frac{1.5 \text{ mm}}{47 \text{ ocular units}} = 0.0319 \text{ mm/OU} = 32\mu m/OU$$

As shown in the table above, it is customary to record more than one measurement. Each measurement is calculated separately. If the calculated ocular unit values differ, use their arithmetic mean as the calibration for that objective lens.

As mentioned earlier, when you change to a higher magnification, the calibration changes. This means you must calibrate for *each* objective lens as before. The only one you can not calibrate directly is the oil immersion lens due to its length and the thickness of the stage micrometer. Its calibration must be calculated from the others, as shown on the next page.

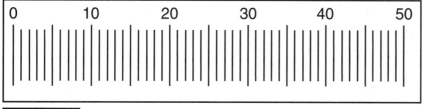

FIGURE 3-2   **An ocular micrometer**

FIGURE 3-3   **A stage micrometer. It is 2.2 mm long, with major divisions of 0.1 mm.**

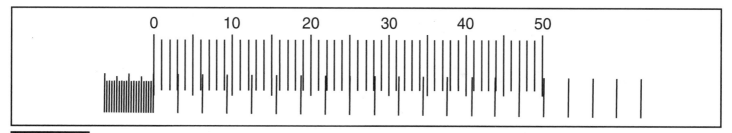

FIGURE 3-4   **The stage micrometer as viewed through the microscope with the ocular micrometer superimposed on it.**

The total magnifications for a typical microscope are given in the table below.

POWER	OBJECTIVE LENS	OCULAR LENS	TOTAL MAGNIFICATION
Low Power	4X	10X	4X × 10X = 40X
Medium Power	10X	10X	10X × 10X = 100X
High Dry Power	40X	10X	40X × 10X = 400X
Oil Immersion	100X	10X	100X × 10X = 1000X

Notice that the magnification from medium power to oil immersion increases by a factor of 10. This means you will see one-tenth as much of the specimen magnified ten-times bigger. Therefore, each ocular unit should also be worth one-tenth as much. Oil immersion can be calculated by dividing the calibration for medium power by 10. Notice also that the high dry calibration can be divided by 2.5 to calculate the oil immersion value.

For practice, assuming the calibration calculated above was for the low power objective, calculate the medium, high dry and oil immersion calibrations. Fill in the table below. (Remember, the calculated size of an ocular unit (OU) on low power was 32 µm.)

TOTAL MAGNIFICATION	CALIBRATION (µm/OU)
40X	32 µm/OU
100X	
400X	
100X	

MATERIALS

Compound microscope equipped with an ocular micrometer
Stage micrometer

INSTRUCTIONS

1. Check your microscope and determine which ocular has the micrometer in it.

2. Move the low power objective into position.

3. Place the stage micrometer on the stage and position it so that its image is superimposed by the ocular micrometer and their lefthand marks line up.

4. Examine the two micrometers and record two or three points where they line up exactly. Record these values in the table below and calculate the value of each ocular unit.

STAGE MICROMETER	OCULAR MICROMETER	CALIBRATION (µm/OU)

5. Change to medium power and repeat the process.

STAGE MICROMETER	OCULAR MICROMETER	CALIBRATION (µm/OU)

6. Change to high dry power and repeat the process.

STAGE MICROMETER	OCULAR MICROMETER	CALIBRATION (μm/OU)

7. Compute average calibrations for each objective lens and record in the table below. Using one of the calibrations, compute the calibration for the oil immersion lens.

OBJECTIVE LENS	AVERAGE CALIBRATION (μm/OU)
Low Power	
Medium Power	
High Dry Power	
Oil Immersion	

8. These calibrations remain valid for this microscope. There is no need to recalibrate unless you change microscopes.

REFERENCES

Baron, Ellen Jo, Lance R. Peterson, and Sydney M. Finegold. 1994. Chapter 8 in *Bailey and Scott's Diagnostic Microbiology, 9th Ed*. Mosby-Year Book, Inc. St. Louis, MO 63146.

Chapin, Kimberle. 1995. Chapter 4 in *Manual of Clinical Microbiology, 6th Ed*., edited by Patrick R. Murray, Ellen Jo Baron, Michael A. Pfaller, Fred C. Tenover, and Robert H. Yolken. American Society for Microbiology, Washington, D.C.

Murray, R.G.E. and Carl F. Robinow. 1994. Chapter 1 in *Methods for General and Molecular Bacteriology*, edited by Philipp Gerhardt, R.G.E. Murray, Willis A. Wood, and Noel R. Krieg. American Society for Microbiology, Washington, D.C.

SIMPLE BACTERIOLOGICAL STAINS

Cytoplasm is transparent and, as such, makes viewing cells with the light microscope difficult without the aid of stains. Staining makes cells more visible so that size, shape and arrangement may be determined more easily. In this set of exercises, you will learn how to correctly prepare a bacterial smear for staining and how to perform simple and negative stains.

Exercise 3–3

Preparing a Bacterial Smear

MATERIALS

Bacterial culture (as assigned)
Clean glass microscope slides
Inoculating loop or needle

PROTOCOL

1. A bacterial smear is made prior to most staining procedures. Figures 3-5a through 3-5d illustrate the procedure for preparing a bacterial smear.

2. Follow with the appropriate staining procedure.

PRECAUTIONS

⚠ Do not use too much water in preparing the slide. This will prolong the air drying step.

⚠ Do not over-inoculate the smear. When dry, it should be barely visible as a film on the glass.

⚠ Although air drying is the most time-consuming step in the procedure, resist the temptation to wave the slide, blow on it, or heat it to speed up the drying process, as contaminating aerosols may result.

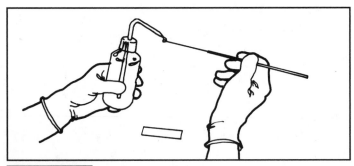

FIGURE 3-5a   Begin With Water
Capture a drop of water with your inoculating loop. If your specimen is growing in broth, you may omit the water drop and continue with step 3-5c.

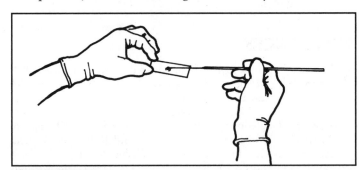

FIGURE 3-5b   Place the Water on the Slide
Transfer the water drop to the center of a clean slide. Avoid using too much water.

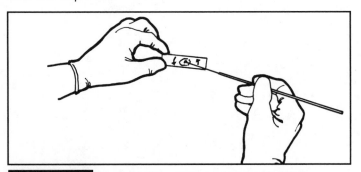

FIGURE 3-5c   Transfer the Organisms to the Slide
Use your loop to aseptically transfer the cells to the water drop. (Avoid excessive inoculation as thick smears are more difficult to stain consistently and individual cells may be difficult to find.) Then, without "springing" your loop, gently *emulsify* (mix) the cells in the drop. As you do so, spread the drop out over the slide's surface so it will air dry more quickly. The slide must be completely dry before continuing.

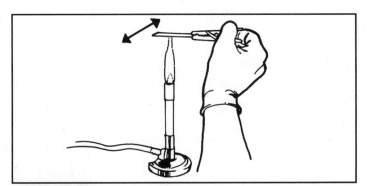

FIGURE 3-5d   Heat-Fix the Slide
Once the drop has air dried, use a slide holder and pass the slide through the upper part of a flame two or three times to heat-fix the smear. Heat-fixing the dried smear makes the cells adhere to the slide, kills them, and makes them more easily stained as protein becomes coagulated. Do not overheat the slide as aerosols may be produced.

⚠ Do not overheat the slide while heat-fixing it. You should be able to handle a properly heat-fixed slide without burning yourself.

REFERENCES

Chapin, Kimberle. 1995. Chapter 4 in *Manual of Clinical Microbiology, 6th Ed.*, edited by Patrick R. Murray, Ellen Jo Baron, Michael A. Pfaller, Fred C. Tenover, and Robert H. Yolken. American Society for Microbiology, Washington, D.C.

Claus, G. William. 1989. Chapter 5 in *Understanding Microbes – A Laboratory Textbook for Microbiology*. W.H. Freeman and Company, New York, NY.

DIFCO Laboratories. 1984. *DIFCO Manual, 10th Ed.* DIFCO Laboratories, Detroit, MI.

Murray, R.G.E., Raymond N. Doetsch and C.F. Robinow. 1994. Page 27 in *Methods for General and Molecular Bacteriology,* edited by Philipp Gerhardt, R.G.E. Murray, Willis A. Wood, and Noel R. Krieg. American Society for Microbiology, Washington, D.C.

Norris, J. R. and Helen Swain. 1971. Chapter II in *Methods in Microbiology, Volume 5A*, edited by J. R. Norris and D. W. Ribbons. Academic Press, Ltd., London.

Power, David A. and Peggy J. McCuen. 1988. Page 4 in *Manual of BBL® Products and Laboratory Procedures, 6th Ed.* Becton Dickinson Microbiology Systems, Cockeysville, MD.

Simple Stain

MATERIALS

Clean glass microscope slides
Methylene blue stain
Safranin stain
Crystal violet stain
Disposable latex gloves
Staining tray
Staining screen
Bibulous paper tablet
Slide holder
Recommended organisms:
  Micrococcus luteus
  Staphylococcus epidermidis
  Bacillus cereus
  Rhodospirillum rubrum
  Vibrio harveyi
  Moraxella catarrhalis

*Photographic Atlas
Reference
Page 21*

STAINING PROTOCOLS

1. Prepare and heat-fix a smear of each organism as described in Figure 3-5.

2. Follow the basic staining procedure illustrated in Figures 3-6a through 3-6c. Prepare two slides with each stain using the following times:

   Crystal violet: stain for 30 to 60 seconds
   Safranin: stain for up to 1 minute
   Methylene blue: stain for 30 to 60 seconds

   (Record the duration of staining in the table below so you can adjust staining time if the specimen is overstained or understained.)

3. Observe using the oil immersion lens. Record your observations of cell morphology, arrangement and size in the table provided.

ORGANISM	STAIN AND DURATION	CELLULAR MORPHOLOGY AND ARRANGEMENT	CELL DIMENSIONS

FIGURE 3-6a **Flood the Smear with Stain**

Place your slide with its smear on the staining rack. Flood the smear with stain for the correct length of time. Hold the slide with a slide holder to minimize staining of your hands. Wearing latex gloves is also a good idea.

FIGURE 3-6b **Rinse with Water**

Tilt the slide to an angle of 45°. Direct a stream of water towards the top of the slide and allow the water to run down across the smear's surface. Continue washing until the runoff is clear.

FIGURE 3-6c **Blot Dry**

Blot (do not wipe) the slide dry in a tablet of bibulous paper. Then observe the specimen using the oil immersion lens.

PRECAUTIONS

⚠ Remember that consistency in preparation produces consistent results. Once you have learned proper smear technique and cell density to achieve optimum staining, stick with it.

⚠ When rinsing the slide, avoid spraying the water directly on the smear as this may dislodge your specimens.

⚠ Dispose of the specimen slides in a jar of disinfectant after use.

REFERENCES

Murray, R.G.E., Raymond N. Doetsch and C.F. Robinow. 1994. Page 27 in *Methods for General and Molecular Bacteriology*, edited by Philipp Gerhardt, R.G.E. Murray, Willis A. Wood, and Noel R. Krieg. American Society for Microbiology, Washington, D.C.

Norris, J. R. and Helen Swain. 1971. Chapter II in *Methods in Microbiology, Volume 5A*, edited by J. R. Norris and D. W. Ribbons. Academic Press, Ltd., London.

Negative Stain

MATERIALS

Nigrosin stain
Clean glass microscope slides
Disposable latex gloves
Recommended organisms:
   Micrococcus luteus
   Bacillus megaterium

STAINING PROTOCOL

1. Figures 3-7a through 3-7e illustrate the procedure for preparing a negative stain.

2. Observe using the oil immersion lens. Record your results in the table on the following page.

Photographic Atlas
Reference
Page 22

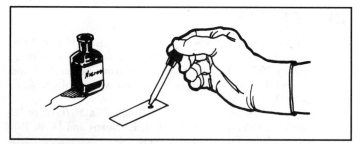

FIGURE 3-7a   **Begin with the Stain**
Place a small drop of nigrosin stain at one end of a clean glass slide. Avoid excess nigrosin on the slide. It is advisable to wear latex gloves to protect your hands.

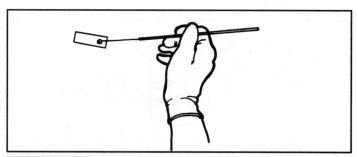

FIGURE 3-7b   **Add the Organism**
Use a loop to aseptically transfer cells to the slide. Gently mix the organisms into the nigrosin. Avoid over-inoculating the slide or spattering the contaminated nigrosin drop as you mix. Flame the loop before proceeding.

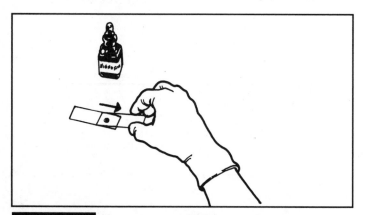

FIGURE 3-7c   **Use a Second Slide as a Spreader**
Place a clean second slide on the surface of the first slide and carefully back it up into the drop of nigrosin.

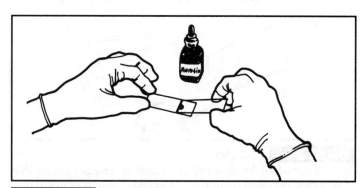

FIGURE 3-7d   **Not Too Far!**
As soon as the nigrosin flows across the width of the spreader slide, stop.

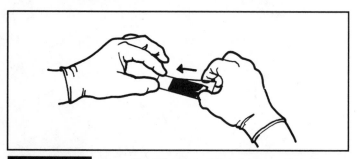

FIGURE 3-7e   **Spread a Nigrosin Film Across the Slide**
Make a nigrosin smear by pushing the spreader slide across the specimen slide's surface. Allow the film to air dry. Dispose of the spreader slide appropriately since it is contaminated.

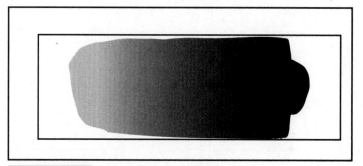

FIGURE 3-7f   **The Finished Product**
After air drying, the slide should look like this. Use the oil lens to observe a region where the nigrosin is relatively thin.

ORGANISM	CELLULAR MORPHOLOGY AND ARRANGEMENT	CELL DIMENSIONS

PRECAUTIONS

⚠ Spread the nigrosin as soon as it flows across the width of the spreader slide. Too much nigrosin in the smear makes microscopic observation difficult.

⚠ Dispose of the spreader slide in a jar of disinfectant immediately after use.

⚠ Dispose of the specimen slide in a jar of disinfectant after use.

REFERENCES

Claus, G. William. 1989. Chapter 5 in *Understanding Microbes – A Laboratory Textbook for Microbiology*. W.H. Freeman and Company, New York, NY.

Murray, R.G.E., Raymond N. Doetsch and C.F. Robinow. 1994. Page 27 in *Methods for General and Molecular Bacteriology*, edited by Philipp Gerhardt, R.G.E. Murray, Willis A. Wood, and Noel R. Krieg. American Society for Microbiology, Washington, D.C.

DIFFERENTIAL STAINS

Differential stains allow a microbiologist to detect differences between organisms or differences between parts of the same organism. In practice, these are used much more frequently than simple stains because they allow determination of cell size, shape and arrangement (as with a simple stain), but also provide information about other features as well. The Gram stain is the most commonly used differential stain in bacteriology. Other differential stains are used for organisms not distinguishable by the Gram stain or for those possessing other important cellular attributes, such as acid-fastness, a capsule, spores, or flagella.

Exercise 3-6 — Gram Stain

MATERIALS

Clean glass microscope slides
Sterile cotton swab
Gram crystal violet
Gram iodine
90% ethanol
Gram safranin
Bibulous paper
Disposable latex gloves
3% KOH
Staining tray
Staining screen
Bibulous paper
Slide holder
Recommended organisms (grown on agar slants):
   Bacillus cereus
   Escherichia coli

STAINING PROTOCOL

1. Prepare and heat-fix a smear of each organism next to one another on the same clean glass slide. Since Gram stains require much practice, you may want to prepare several slides and let them be air drying simultaneously. Then they'll be ready if you need them.

2. Use the sterile swab to obtain a sample from your teeth at the gum line. Transfer the sample to a drop of water on a clean glass slide, air dry and heat fix.

3. Follow the basic staining procedure illustrated in Figures 3-8a through 3-8i. We recommend staining the pure cultures first. When your technique is consistent, then stain the tooth sample.

Photographic Atlas Reference Page 27

4. Observe using the oil immersion lens. Record your observations of cell morphology and arrangement, dimensions, and Gram reactions in the table provided.

Protocol for Determining Gram Reaction Using KOH

1. Make an emulsion of one Gram-positive and one Gram-negative organism (as determined by your Gram stain) on the same slide.

2. Add a loopful of 3% KOH to one smear and mix for a minute.

3. After mixing, remove the loop slowly.

4. Repeat with the other emulsion.

5. Emulsions of Gram-negative cells become viscous and adhere to the loop. As the loop is lifted off the slide a thin strand of the mucoid mixture will stretch between them. This property is not seen with Gram-positive organisms.

ORGANISM	CELLULAR MORPHOLOGY AND ARRANGEMENT	CELL DIMENSIONS	GRAM REACTION (+ / −)

FIGURE 3-8a   **Flood the Smear with Crystal Violet**

Place your slide with its smears on the staining rack. Flood the smear with crystal violet and let it stand for one minute. Hold the slide with a slide holder to minimize staining of your hands. Wearing latex gloves is also a good idea.

FIGURE 3-8b   **Rinse with Water**

Tilt the slide and rinse away excess crystal violet by directing a stream of water towards the top of the slide and allowing the water to run down across the smear's surface. Avoid spraying the water directly on the smear as this may dislodge your specimens.

FIGURE 3-8c   **Flood the Smear with Iodine**

Cover the smear with iodine solution for at least one minute.

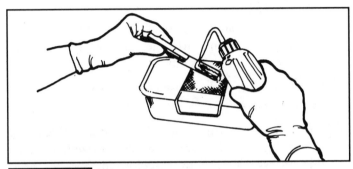

FIGURE 3-8d   **Rinse Again with Water**

Gently wash off the excess iodine as in Figure 3-8b.

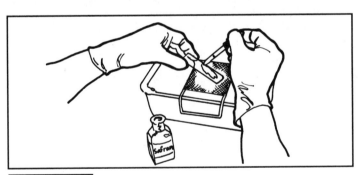

FIGURE 3-8e   **Decolorize**

Tilt the slide and allow the alcohol to run across the smear. Stop decolorizing when the run-off is clear, but no longer than 30 seconds.

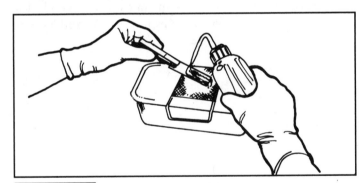

FIGURE 3-8f   **Rinse Again with Water**

Immediately rinse off the alcohol with water as in Figure 3-8b. The longer you delay, the greater the likelihood of overdecolorizing.

FIGURE 3-8g   **Counterstain with Safranin**

Cover the smear with safranin for one minute.

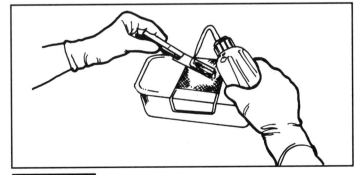

FIGURE 3-8h   **Rinse Again with Water**

Gently wash off the excess safranin as in Figure 3-8b.

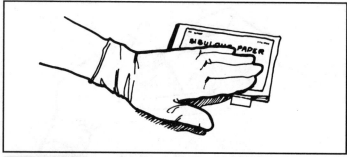

FIGURE 3-8i **Blot Dry**
Blot (do not wipe) the slide dry in a tablet of bibulous paper. Then observe the specimen using the oil immersion lens.

PRECAUTIONS

⚠ Some Gram-positive microbes lose their ability to resist decolorization with age. Always use cultures younger than 24 hours old.

⚠ Decolorization is the most critical step in the procedure.

⚠ Strive for preparing smears of uniform thickness. Thick smears risk being underdecolorized, whereas thin smears risk being overdecolorized.

⚠ Until you are confident of your ability to produce consistent and reliable Gram stains, it is a good idea to run controls (known Gram-positive and Gram-

negative organisms) next to your specimen (see Fig. 4-3 in the *Photographic Atlas*).

⚠ Dispose of the specimen slides in a jar of disinfectant after use.

REFERENCES

Baron, Ellen Jo, Lance R. Peterson, and Sydney M. Finegold. 1994. Chapter 10 in *Bailey and Scott's Diagnostic Microbiology, 9th Ed.* Mosby-Year Book, Inc. St. Louis, MO 63146.

Chapin, Kimberle. 1995. Chapter 4 in *Manual of Clinical Microbiology, 6th Ed.*, edited by Patrick R. Murray, Ellen Jo Baron, Michael A. Pfaller, Fred C. Tenover, and Robert H. Yolken. American Society for Microbiology, Washington, D.C.

Koneman, Elmer W., Stephen D. Allen, William M. Janda, Paul C. Schreckenberger and Washington C. Winn, Jr. 1997. Chapter 14 in *Color Atlas and Textbook of Diagnostic Microbiology, 5th Ed.* J.B. Lippincott Company, Philadelphia, PA.

Murray, Patrick R., Ellen Jo Baron, Michael A. Pfaller, Fred C. Tenover, and Robert H. Yolken. 1995. *Manual of Clinical Microbiology, 6th Ed.* American Society for Microbiology, Washington, D.C.

Murray, R.G.E., Raymond N. Doetsch and C.F. Robinow. 1994. Pages 31 and 32 in *Methods for General and Molecular Bacteriology*, edited by Philipp Gerhardt, R.G.E. Murray, Willis A. Wood, and Noel R. Krieg. American Society for Microbiology, Washington, D.C.

Norris, J. R. and Helen Swain. 1971. Chapter II in *Methods in Microbiology, Volume 5A*, edited by J. R. Norris and D. W. Ribbons. Academic Press, Ltd., London.

Power, David A. and Peggy J. McCuen. 1988. Page 261 in *Manual of BBL® Products and Laboratory Procedures, 6th Ed.* Becton Dickinson Microbiology Systems, Cockeysville, MD.

Acid-Fast Stains

Exercise 3–7

MATERIALS

Clean glass microscope slides
Staining tray
Staining screen
Bibulous paper
Slide holder
Ziehl-Neelsen Stains
   Methylene blue stain
   Ziehl's carbolfuchsin stain
   Acid alcohol
Kinyoun Stains
   Kinyoun carbolfuchsin
   Acid alcohol
   Brilliant green stain
Sheep serum
Heating apparatus (steam or hot plate)
Disposable latex gloves
Lab coat or apron
Eye goggles
Recommended organisms:
   Mycobacterium phlei
   Bacillus subtilis

STAINING PROTOCOL — ZIEHL-NEELSEN (ZN) METHOD

1. Prepare smears of each organism on a clean glass slide, substituting a drop of sheep serum for the drop of water. Air dry, then heat-fix the smears. NOTE: you may make two separate smears right next to one another on the slide or mix the two organisms in one smear.

2. Cover the slide with a strip of bibulous paper and place it on the heating apparatus. The paper strip should be the same size as the slide.

3. Working in a fume hood, saturate the paper with Ziehl's carbolfuchsin and keep it moist as you steam

the slide for 5 minutes (Fig. 3-9). It is advisable to wear latex gloves while staining. A lab coat or apron and eye goggles should

> **Photographic Atlas
> Reference
> Page 29**

protect you from spattering stain in case the slide overheats. Be sure to clean up any spills.

4. After 5 minutes of *steaming*, remove the paper and rinse off the excess carbolfuchsin with water. Hold the slide on an angle. Aim the water stream above the bacterial smear and allow the water to run across the smear (Fig. 3-6b).

5. Decolorize with acid alcohol until the runoff is clear. Caution! Do not hold the slide with your fingers. This is an acidic solution.

6. Rinse with water as before.

7. Counterstain with methylene blue for one minute (Fig. 3-6a).

8. Rinse with water, then blot dry with bibulous paper.

9. Observe using the oil immersion lens. Record your observations of cell morphology and arrangement, dimensions, color, and acid-fast reaction in the space provided.

STAINING PROTOCOL — KINYOUN METHOD

1. Prepare smears of each organism on a clean glass slide, substituting a drop of sheep serum for the drop of water. Air dry, then heat-fix the smears. NOTE: you may make two separate smears right next to one another on the slide or mix the two organisms in one smear.

2. Flood the slide with Kinyoun carbolfuchsin stain for 5 minutes (Fig. 3-6a).

ORGANISM	CELLULAR MORPHOLOGY AND ARRANGEMENT	CELL DIMENSIONS	COLOR	ACID FAST REACTION (+/−)

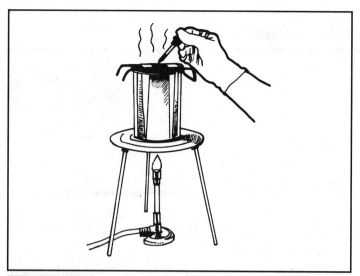

FIGURE 3-9 **Steaming the Slide During the Ziehl-Neelsen Procedure**
Carefully steam the slide to force the carbolfuchsin into acid-fast cells. Do not boil the slide or let it dry out. Keep it moist with stain for the entire five minutes of steaming. This should be performed in a well-ventilated area with hand, clothing and eye protection.

3. After 5 minutes of staining, rinse off the excess carbolfuchsin with water. Hold the slide on a 45° angle. Aim the water stream above the bacterial smear and allow the water to run across the smear (Fig. 3-6b).

4. Decolorize with acid alcohol until the runoff is clear (Fig. 3-8e). Caution! Do not hold the slide with your fingers. This is an acidic solution.

5. Rinse with water as before (Fig. 3-6b).

6. Counterstain with brilliant green for 30 seconds (Fig. 3-6a).

7. Rinse with water as before, then blot dry with bibulous paper (Fig. 3-6c).

8. Observe using the oil immersion lens. Record your observations of cell morphology and arrangement, dimensions, color, and acid-fast reaction in the space provided.

PRECAUTIONS

⚠ Perform both stains in a well-ventilated area.

⚠ Decolorization is not nearly as challenging in acid-fast staining as in the Gram stain, but uniform smear preparation is still necessary to achieve consistent results.

⚠ Use the acid alcohol with caution.

⚠ Keep the paper moist with stain while steaming it during the ZN procedure.

⚠ Do not boil the stain on the slide during the ZN procedure. This may cause spattering of stain and crack the slide.

REFERENCES

Chapin, Kimberle. 1995. Chapter 4 in *Manual of Clinical Microbiology, 6th Ed.*, edited by Patrick R. Murray, Ellen Jo Baron, Michael A. Pfaller, Fred C. Tenover, and Robert H. Yolken. American Society for Microbiology, Washington, D.C.

Murray, Patrick R., Ellen Jo Baron, Michael A. Pfaller, Fred C. Tenover, and Robert H. Yolken. 1995. *Manual of Clinical Microbiology, 6th Ed.* American Society for Microbiology, Washington, D.C.

Murray, R.G.E., Raymond N. Doetsch and C.F. Robinow. 1994. Page 32 in *Methods for General and Molecular Bacteriology,* edited by Philipp Gerhardt, R.G.E. Murray, Willis A. Wood, and Noel R. Krieg. American Society for Microbiology, Washington, D.C.

Norris, J. R. and Helen Swain. 1971. Chapter II in *Methods in Microbiology, Volume 5A*, edited by J. R. Norris and D. W. Ribbons. Academic Press, Ltd., London.

Power, David A. and Peggy J. McCuen. 1988. Page 5 in *Manual of BBL® Products and Laboratory Procedures, 6th Ed.* Becton Dickinson Microbiology Systems, Cockeysville, MD.

ORGANISM	CELLULAR MORPHOLOGY AND ARRANGEMENT	CELL DIMENSIONS	COLOR	ACID FAST REACTION (+/−)

Capsule Stain

MATERIALS

Clean glass slides
Sheep serum
Maneval's stain
Congo red stain
Staining tray
Staining screen
Bibulous paper tablet
Slide holder
Bibulous paper
Latex gloves
Recommended organisms:
  18 to 24 hour skim milk pure culture of *Flavobacterium capsulatum (Sphingomonas capsulata)*
  18 to 24 hour nutrient agar slant pure culture of *Bacillus subtilis*

STAINING PROTOCOL

Each specimen should be done on a separate slide.

1. Place a small drop of sheep serum at one end of a clean glass slide. Add a drop of Congo red stain, then aseptically transfer a small amount of *F. capsulatum* to the drop and mix. Wearing latex gloves to protect your hands is a good idea. Be sure to flame your inoculating loop after the transfer.

2. Use a clean slide to spread a thin film of the serum/stain/bacteria drop as in the negative stain (Figs. 3-7c through 3-7e). Be sure to dispose of the spreader slide properly.

3. Allow the slide to air dry completely.

4. Flood the slide with Maneval's stain for one minute (Fig. 3-6a).

5. Rinse with water (Fig. 3-6b), then blot dry with bibulous paper.

Photographic Atlas Reference Page 31

6. Observe using the oil immersion lens. Record your observations of cell morphology and arrangement, cell dimensions, and presence or absence of a capsule in the table provided.

PRECAUTIONS

⚠ Be sure to dispose of the spreader slides properly in disinfectant.

⚠ Be sure to sterilize the inoculating loop after the transfer.

⚠ Do not heat-fix the slides. This will produce artifactual white halos around the cells that may be interpreted as capsules.

⚠ Dispose of the specimen slide in a jar of disinfectant after use.

REFERENCES

Murray, R.G.E., Raymond N. Doetsch and C.F. Robinow. 1994. Page 35 in *Methods for General and Molecular Bacteriology*, edited by Philipp Gerhardt, R.G.E. Murray, Willis A. Wood, and Noel R. Krieg. American Society for Microbiology, Washington, D.C.

Norris, J. R. and Helen Swain. 1971. Chapter II in *Methods in Microbiology, Volume 5A*, edited by J. R. Norris and D. W. Ribbons. Academic Press, Ltd, London.

ORGANISM	CELLULAR MORPHOLOGY AND ARRANGEMENT	CELL DIMENSIONS	CAPSULE (+/−)

**Exercise
3–9**

Spore Stain

MATERIALS

*Photographic Atlas
Reference
Page 32*

Clean glass microscope slides
Malachite green stain
Safranin stain
Heating apparatus (steam apparatus or hot plate)
Bibulous paper
Staining tray
Staining screen
Slide holder
Disposable latex gloves
Lab coat or apron
Goggles
Recommended organisms:
   18 to 24 hour nutrient agar slant pure culture of
   Staphylococcus epidermidis
   5 day nutrient agar slant pure culture of *Bacillus cereus*

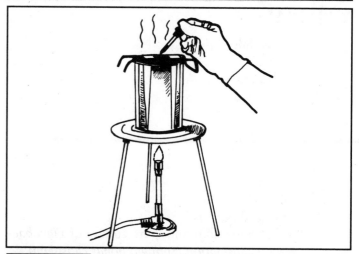

FIGURE 3-10   **Steaming the Slide**
Carefully steam the slide to force the malachite green into the spores. Do not boil the slide or let it dry out. Keep it moist with stain for the entire five minutes. This should be done in a well-ventilated area or inside a fume hood with hand, clothing, and eye protection.

STAINING PROTOCOL

1. Prepare and heat-fix a smear of each organism on the same slide.

2. Cover the slide with a strip of bibulous paper and place it on the heating apparatus. The paper strip should be the same size as the slide.

3. Working in a fume hood, saturate the paper with malachite green stain and steam the slide for 5 minutes. Do not allow the paper to dry: add more stain as needed to keep it moist (Fig. 3-10). It is advisable to wear latex gloves while staining. A lab coat or apron and eye goggles should protect you from spattering stain in case the slide overheats. Be sure to clean up any spills.

4. After 5 minutes of *steaming*, remove the paper and gently rinse off the excess malachite green with water. Water also acts as the decolorizing agent in this procedure, so rinse thoroughly (Fig. 3-6b).

5. Counterstain with safranin for up to one minute (Fig. 3-6a).

6. Rinse gently with water, then blot dry with bibulous paper (Figs. 3-6b and 3-6c).

7. Observe using the oil immersion lens. Record your observations of cell morphology and arrangement, cell dimensions, and spore presence, position and shape in the table provided.

ORGANISM	CELLULAR MORPHOLOGY AND ARRANGEMENT	CELL DIMENSIONS	SPORES (+/−)	SPORE SHAPE	SPORE POSITION

PRECAUTIONS

⚠ Perform this stain in a well-ventilated area, preferably in a fume hood.

⚠ Keep the paper moist with stain while steaming it.

⚠ Do not boil the stain on the slide. This may cause spattering of stain and crack the slide. It may also destroy any vegetative cells present.

⚠ Absence of spores does not necessarily mean the organism *cannot* produce them; cultures that are too young may not have had sufficient time to produce them.

⚠ Dispose of the specimen slide in a jar of disinfectant after use.

REFERENCES

Claus, G. William. 1989. Chapter 9 in *Understanding Microbes – A Laboratory Textbook for Microbiology*. W.H. Freeman and Company, New York, NY.

Murray, R.G.E., Raymond N. Doetsch and C.F. Robinow. 1994. Page 34 in *Methods for General and Molecular Bacteriology*, edited by Philipp Gerhardt, R.G.E. Murray, Willis A. Wood, and Noel R. Krieg. American Society for Microbiology, Washington, D.C.

Flagella Stain

MATERIALS

Sterile nutrient broth
Pasteur pipettes
Centrifuge and centrifuge tubes
10% Formalin
Mildly abrasive cleanser
Wax china marker
Pararosaniline stain
Microscope slides
Staining tray
Staining screen
Bibulous paper tablet
Slide holder
Commercially prepared slides of motile organisms
Recommended organisms (grown on agar slants):
  Proteus vulgaris
  Staphylococcus epidermidis

STAINING PROTOCOL

Each specimen should be done on a separate slide.

1. Clean a slide with a thin solution of household cleanser, such as Bon Ami. Allow the paste to dry in a thin film on the slide, then remove thoroughly with a tissue.

2. Add a few milliliters of nutrient broth to the agar slant with a Pasteur pipette.

3. After a few minutes, decant the broth into a centrifuge tube.

4. Centrifuge (longer is better than faster) until a visible pellet is seen. Decant the supernatant from the tube and resuspend the pellet in 10% (v/v) Formalin. In all of these steps, handle the bacteria gently, as the flagella are fragile and may break off.

5. Transfer a loop of the bacterial suspension to a cleaned slide. Hold the slide on an angle and allow the drop to run down it. Air dry the bacterial film.

6. Draw a rectangle around the film using a china marker.

7. Flood the rectangular region with the pararosaniline stain until a golden film with precipitate forms (visible with transmitted light) throughout the preparation. This may take up to 15 minutes.

8. Rinse gently with water, then air dry.

9. Observe using the oil immersion lens. Record your observations of cell morphology, arrangement, and dimensions, and flagellar presence and arrangement in the table provided.

10. Commercially prepared slides are available and provide satisfactory material for observation. Record your observations in the spaces below.

> **Photographic Atlas Reference Page 34**

PRECAUTIONS

⚠ To minimize the effects of stain precipitates and other artifacts, only use slides that have been cleaned and degreased.

⚠ Bacterial flagella are extremely fragile. Handle the preparations very gently to minimize the inevitable damage.

⚠ Flagella are very small and are not easily seen. Use the oil lens and patiently scan your preparation until you find a field with satisfactory specimens. Even commercially prepared slides may require more than the usual effort to find good examples.

⚠ Dispose of your specimen slide in a jar of disinfectant after use.

REFERENCES

Iino, Tetsuo and Masatoshi Enomoto. 1971. Chapter IV in *Methods in Microbiology, Volume 5A*, edited by J. R. Norris and D. W. Ribbins. Academic Press, Ltd., London.

Murray, R.G.E., Raymond N. Doetsch and C.F. Robinow. 1994. Page 35 in *Methods for General and Molecular Bacteriology*, edited by Philipp Gerhardt, R.G.E. Murray, Willis A. Wood, and Noel R. Krieg. American Society for Microbiology, Washington, D.C.

ORGANISM	CELLULAR MORPHOLOGY AND ARRANGEMENT	CELL DIMENSIONS	FLAGELLA (+ / −)	FLAGELLAR ARRANGEMENT

Exercise 3-11

Hanging Drop and Wet Mount Preparation

MATERIALS

Clean glass slides
Depression slide and cover glass
Petroleum jelly
Toothpick
Recommended organisms (grown on solid media):
    Proteus vulgaris
    Staphylococcus epidermidis

PROTOCOL — HANGING DROP PREPARATION (FOR LONG-TERM OBSERVATION)

> *Photographic Atlas Reference Page 35*

1. Follow the procedure illustrated in Figures 3-11a through 3-11d for each specimen.

2. Observe under high dry or oil immersion and record your results in the table below.

3. When finished, remove the cover glass. Soak it and the slide in a jar of disinfectant for 15 minutes.

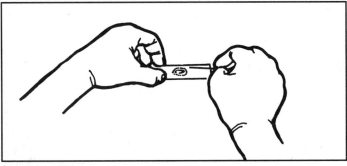

FIGURE 3-11a   **Apply the Petroleum Jelly**
Use a toothpick to place a *thin* ring of petroleum jelly around the well of a depression slide.

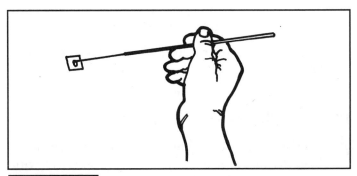

FIGURE 3-11b   **Transfer the Organisms**
Use your loop to place a drop of water on a cover glass. Then, aseptically transfer the bacteria to the drop. Do not emulsify or spread out the drop. Flame your loop immediately after the transfer.

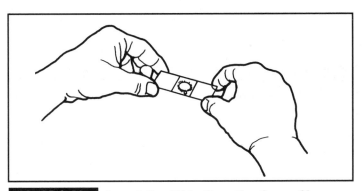

FIGURE 3-11c   **Invert the Slide Over the Cover Glass**
Carefully invert the depression slide over the cover glass so the drop is centered in its well. Gently press until the petroleum jelly has created a seal between the slide and cover glass.

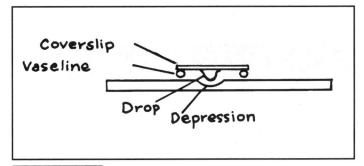

FIGURE 3-11d   **Observe the Slide**
Correctly done, the slide should look like this from the side. Place the slide with the cover glass up on the microscope stage and observe under high dry or oil immersion.

ORGANISM	CELLULAR MORPHOLOGY AND ARRANGEMENT	CELL DIMENSIONS	MOTILITY (+/−)

4. After 15 minutes in the disinfectant, remove the depression slide, rinse it with water and clean with ethanol to remove the petroleum jelly. Rinse with water and dry. The slide may be used again.

PROTOCOL — WET MOUNT PREPARATION (FOR SHORT-TERM OBSERVATION)

1. Place a drop of water on a clean glass slide.

2. Add bacteria to the drop. Don't over-inoculate.

3. Gently lower a cover glass with your loop supporting one side over the drop of water. Avoid trapping air bubbles (Fig. 10-2).

4. Observe under high dry or oil immersion and record your results in the table below.

5. Dispose of the wet mount slide and cover glass in a sharps container.

PRECAUTIONS

⚠ Avoid excessive petroleum jelly on your hanging drop slide. Petroleum jelly in the well will interfere with your observations. If the layer is too thick, your slide will not fit under the oil immersion lens.

⚠ Be careful not to squash your hanging drop preparation with the high dry or oil immersion lenses while focusing. There is also the risk of the oil lens pushing the cover glass out of the way as you swing it into place.

⚠ Begin observation at the edge of the drop. This will make focusing easier and may provide a better view of motile microbes if they are aerobic.

⚠ In both preparations, be sure to distinguish between true motility and Brownian motion. Brownian motion is not true motility but vibration due to the kinetic energy of water molecules crashing into the cells. If the cells are not vibrating at all, they may be stuck on the glass slide.

⚠ Be aware of water currents in the drop that may give the appearance of motility. (This is especially true in wet mounts. As the water dries, the cells are "herded" in one direction. This is *not* motility.)

REFERENCES

Iino, Tetsuo and Masatoshi Enomoto. 1969. Chapter IV in *Methods in Microbiology, Volume 1,* edited by J. R. Norris and D. W. Ribbins. Academic Press, Ltd., London.

Murray, R.G.E., Raymond N. Doetsch and C.F. Robinow. 1994. Page 26 in *Methods for General and Molecular Bacteriology,* edited by Philipp Gerhardt, R.G.E. Murray, Willis A. Wood, and Noel R. Krieg. American Society for Microbiology, Washington, D.C.

Quesnel, Louis B. 1969. Chapter X in *Methods in Microbiology, Volume 1,* edited by J. R. Norris and D. W. Ribbins. Academic Press, Ltd., London.

ORGANISM	CELLULAR MORPHOLOGY AND ARRANGEMENT	CELL DIMENSIONS	MOTILITY (+/−)

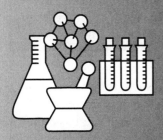

Isolation Techniques and Selective Media

*I*ndividual microbial species from in a mixed culture must be isolated before they can be correctly identified. Several methods for achieving isolation are available, but the simplest and most basic is the streak plate technique on nutrient agar. More sophisticated methods involve streaking on *selective* or *enrichment* medium. Selective media include ingredients that inhibit growth of undesired organisms while enrichment media include nutrients necessary to cultivate desired organisms and lack nutrients necessary for growth of undesired organisms. Some media are formulated to be selective and *differential*. That is, they favor growth of a particular group of bacteria and also, by means of varying growth characteristics or color production, allow differentiation between species.

In laboratory situations, microbiologists are familiar with the pathogens commonly seen in patient samples and the areas of the body from which they come. Armed with this information, they can choose a selective medium that favors the growth of a specific pathogen and inhibits growth of nonpathogenic organisms also likely to be present at the site of infection. This makes isolation and subsequent identification of the pathogens easier since it "narrows the field" from the start; time is not wasted in identifying bacteria that are part of the normal flora.

This section is divided into three units, each dealing with a particular aspect of isolation technique. The first unit is devoted to simple streak plate technique. The second and third units examine some commonly used selective (and often differential) media for the isolation of Gram-positive cocci and Gram-negative rods. In most real clinical situations these media would be streaked for isolation as described in "Streak Plate Method of Isolation." For our purposes, however, the selective and differential capabilities of these media are sometimes best demonstrated with "spot" inoculations. The test protocols of most exercises in this section typically call for spot inoculations of several organisms on a single plate.

STREAKING FOR MICROBIAL ISOLATION

The streak plate technique is a common method of obtaining microbial isolation from a mixed culture. The principle is simple: a loopful of the mixed culture is streaked across the agar surface. As streaking progresses, the density of cells decreases, eventually leading to separated, individual cells being deposited on the agar surface. These cells grow into individual colonies containing thousands of bacteria which are all clones of the original cell. In most cases these colonies are said to be "pure"; that is, they are composed of only one species of bacteria. As such, they can serve as a pure source of organisms for culturing and testing.

In this first exercise, you will develop the basic motor skills necessary for successful streak plates. In the exercises which follow, you will apply this skill toward isolation of specific bacterial groups using specialized selective and differential media.

Exercise 4–1   Streak Plate Method of Isolation

MATERIALS

Two nutrient agar plates per student

Broth culture of *Chromobacterium violaceum* and *Escherichia coli* mixed together in equal portions immediately before use

Broth culture of *Enterobacter aerogenes* and *Escherichia coli* mixed together in equal portions immediately before use

TEST PROTOCOL

1. The quadrant method of isolation is performed as illustrated in Figures 4-1a through 4-1d. In this exercise, you will be streaking nutrient agar. However, in a clinical laboratory, the streak technique is also used in conjunction with selective media such as the ones that comprise the remainder of this section.

2. Before you attempt to streak on a real agar plate, you may practice the wrist action using a pencil on the "plate" in Figure 4-2.

3. Streak the mixture of *Chromobacterium violaceum* and *Escherichia coli* on one of your nutrient agar plates. Label the plate with your name, the date, and the organisms.

4. Streak the mixture of *Enterobacter aerogenes* and *Escherichia coli* on the other nutrient agar plate. Label the plate with your name, the date, and the organisms.

5. Tape the two plates together and incubate them in an inverted position for at least 24 hours at 37°C.

6. After incubation, observe the plates for isolation. Examine the streak pattern to determine in which streak you achieved isolation. If no isolation was obtained, try to determine if technique was responsible.

PRECAUTIONS

Photographic Atlas
Reference
Page 9

 Avoid cutting into the agar surface with the loop as you streak.

 Flame the loop between streaks and especially when you are finished.

 Be careful not to use a hot loop for streaking as you will create aerosols and may kill the organism you are trying to isolate.

 Be sure to rotate the plate 90° between each streak and pass through the preceding streak pattern, otherwise you'll have no organisms to streak!

 Be aware of the plate's rotation direction during the procedure. If you rotate clockwise the first time and counterclockwise thereafter, you will not likely have a successful streak plate.

 Since isolated colonies is the goal of this procedure, be sure to incubate the plate in an inverted position. If you don't, condensation dropping from the lid will wash across the agar surface and the growth will not be in distinct colonies.

REFERENCES

Baron, Ellen Jo, Lance R. Peterson and Sydney M. Finegold. 1994. Chapter 9 in *Bailey and Scott's Diagnostic Microbiology, 9th Ed.* Mosby-Yearbook, St. Louis, MO.

Collins, C. H. and Patricia M. Lyne. 1995. Chapter 6 in *Collins and Lyne's Microbiological Methods, 7th Ed.* Butterworth-Heineman.

Koneman, Elmer W., Stephen D. Allen, William M. Janda, Paul C. Schreckenberger and Washington C. Winn, Jr. 1997. Chapter 2 in *Color Atlas and Textbook of Diagnostic Microbiology, 5th Ed.* J.B. Lippincott Company, Philadelphia, PA.

Power, David A. and Peggy J. McCuen. 1988. Pages 2 and 3 in *Manual of BBL® Products and Laboratory Procedures, 6th Ed.* Becton Dickinson Microbiology Systems, Cockeysville, MD.

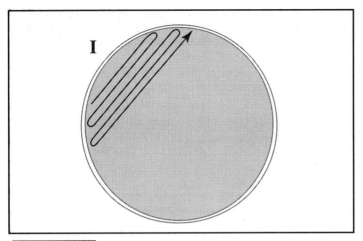

FIGURE 4-1a **Beginning the Streak Pattern**
Streak the mixed culture back and forth in one quadrant of the agar plate. Use the lid as a shield. Flame the loop, then proceed.

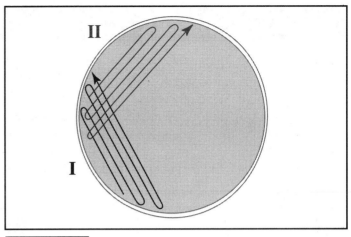

FIGURE 4-1b **Streak Again**
Rotate the plate 90° and touch the agar in an uninoculated region to cool the loop. Streak again using the same wrist motion. Flame the loop.

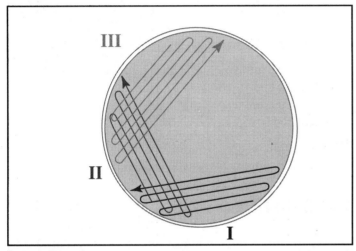

FIGURE 4-1c **Streak Yet Again**
Rotate the plate 90° and streak again using the same wrist motion. Be sure to cool the loop prior to streaking. Flame again.

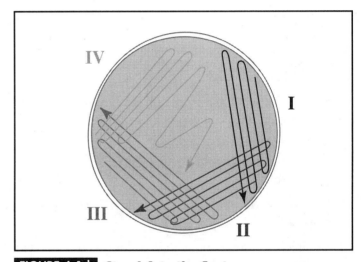

FIGURE 4-1d **Streak Into the Center**
After cooling the loop, streak one last time into the center of the plate. Flame the loop and incubate the plate in an inverted position for the assigned time at the appropriate temperature.

FIGURE 4-2 **A Practice "Plate"**
Practice the streaking technique with a pencil on this "plate" before actually trying it on real media. Keep your wrist relaxed.

SELECTIVE MEDIA FOR ISOLATION OF GRAM-POSITIVE COCCI

Selective media used in the isolation of Gram-positive cocci are used primarily for the identification of pathogenic streptococci and staphylococci. Although blood agar is actually used in the isolation and cultivation of many different bacterial groups, it may be best known for its role in the differentiation of streptococci. In fact, it is so important in this capacity that streptococci are grouped according to their hemolytic properties on blood agar — often determined at the time of isolation. Thus, it is included in this section and not Section Five. Mannitol salt agar is a selective medium which, due to its salt content, specifically favors pathogenic staphylococci. Phenylethyl alcohol agar favors growth of both *Streptococcus* and *Staphylococcus* species and inhibits Gram-negative organisms.

Exercise 4–2

Blood Agar

MATERIALS

One blood agar plate (TSA containing 5% sheep blood)
Sterile cotton swabs

TEST PROTOCOL

1. Have your lab partner obtain a culture from your throat using the technique in Figure 1-2a.

2. Immediately transfer the specimen to a blood agar plate. Use the swab to begin a streak for isolation (Fig. 1-9b).

3. Dispose of the swab in a container designated for autoclaving.

4. Complete the streaking with your loop as described in Figures 4-1b through 4-1d.

5. Label the plate with the your name, the specimen source ("throat culture") and the date.

6. Tape the lid down to prevent it from opening accidentally. Invert and incubate the plate aerobically at 25°C for 24 hours.

7. After incubation, do NOT open your plate unless advised to do so by your instructor.

8. Observe with transmitted and reflected light for color and clearing around the isolated growth. Compare your results with others in your lab.

> *Photographic Atlas*
> *Reference*
> *Page 41*

9. Using the information in the Table of Results, record your results in the space provided. Remember, you streaked a mixed culture from your throat. Don't be surprised if you have a variety of results.

10. If possible, your instructor may have inoculated plates that illustrate the various hemolytic reactions.

PRECAUTIONS

⚠ We recommend buying commercially prepared 5% Sheep Blood Agar plates.

⚠ Blood agar plates must be stored in the refrigerator prior to incubation.

⚠ Since you are isolating from a potentially diseased source (yourself) and don't know what you have isolated, do NOT open your plate after incubation unless your instructor advises you to do so. Examine the plate from underneath to check for hemolysis reactions.

TABLE OF RESULTS		
RESULT	**INTERPRETATION**	**SYMBOL**
Clearing around growth	Organism hemolyzes RBCs completely	β hemolysis
Greening around growth	Organism hemolyzes RBCs partially	∝ hemolysis
No change in the medium	Organism does not hemolyze RBCs	no (γ) hemolysis

SOURCE OF CULTURE	RESULT	INTERPRETATION

REFERENCES

Baron, Ellen Jo, Lance R. Peterson, and Sydney M. Finegold. 1994. Chapter 9 in *Bailey and Scott's Diagnostic Microbiology, 9th Ed*. Mosby-Year Book, Inc. St. Louis, MO 63146.

DIFCO Laboratories. 1984. Page 139 in *DIFCO Manual, 10th Ed*. DIFCO Laboratories, Detroit, MI.

Koneman, Elmer W., Stephen D. Allen, William M. Janda, Paul C. Schreckenberger and Washington C. Winn, Jr. 1997. Chapter 12 in *Color Atlas and Textbook of Diagnostic Microbiology, 5th Ed*. J.B. Lippincott Company, Philadelphia, PA.

Krieg, Noel R. 1994. Page 619 in *Methods for General and Molecular Bacteriology*, edited by Philipp Gerhardt, R. G. E. Murray, Willis A. Wood and Noel R. Krieg, American Society for Microbiology, Washington, DC.

Power, David A. and Peggy J. McCuen. 1988. Page 115 in *Manual of BBL® Products and Laboratory Procedures, 6th Ed*. Becton Dickinson Microbiology Systems, Cockeysville, MD.

Exercise 4–3

Mannitol Salt Agar (MSA)

MATERIALS

One Mannitol Salt Agar plate

Broth culture of *Staphylococcus aureus* and *Staphylococcus epidermidis* mixed together in equal portions immediately before use

TEST PROTOCOL

1. Streak the plate for isolation with the fresh mixture of *S. aureus* and *S. epidermidis*.

2. Label the plate with the organisms' names, your name, and the date.

3. Invert the plate and incubate at 35°C for 24 to 48 hours.

4. Examine the plate and, using the information in the table below, record your results in the space provided.

PRECAUTION

⚠ Use caution examining plates if you streaked from a patient's sample since this medium selects for pathogenic staphylococci.

> *Photographic Atlas Reference Page 15*

REFERENCES

Baron, Ellen Jo, Lance R. Peterson and Sydney M. Finegold. 1994. Chapter 25 in *Bailey and Scott's Diagnostic Microbiology, 9th Ed.* Mosby-Yearbook, St. Louis, MO.

DIFCO Laboratories. 1984. Page 558 in *DIFCO Manual, 10th Ed.* DIFCO Laboratories, Detroit, MI.

Power, David A. and Peggy J. McCuen. 1988. Page 193 in *Manual of BBL® Products and Laboratory Procedures, 6th Ed.* Becton Dickinson Microbiology Systems, Cockeysville, MD.

TABLE OF RESULTS	
RESULT	**INTERPRETATION**
Poor growth or no growth	Organism is inhibited by salt
Good growth	Organism is not inhibited by salt
Yellow growth	Organism produces acid from mannitol fermentation
Red growth	Organism does not ferment mannitol

ORGANISM	RESULT	REMARKS

Exercise 4–4

Phenylethyl Alcohol (PEA) Agar

MATERIALS

One PEA plate
Recommended organisms:
Enterococcus faecalis
Staphylococcus epidermidis
Proteus mirabilis

TEST PROTOCOL

1. Use a marking pen to divide the plate into four sectors. Marks should be made on the plate's base.

2. Spot inoculate (Fig. 1-10a) in different sectors and leave the fourth sector uninoculated as a control.

3. Label the plate with the organisms' names, your name, and the date.

4. Invert the plate and incubate at 35°C for 24 to 48 hours.

5. Examine the plate and, using the information in the Table of Results, record your results in the space provided.

PRECAUTION

⚠ PEA agar is designed to isolate gram-positive microbes, especially gram-positive cocci, by inhibiting gram-negatives. Some gram-negatives, however, are only partially inhibited and may grow slowly on this medium.

Photographic Atlas Reference Page 16

REFERENCES

DIFCO Laboratories. 1984. Page 667 in *DIFCO Manual, 10th Ed.* DIFCO Laboratories, Detroit, MI.
Power, David A. and Peggy J. McCuen. 1988. Page 223 in *Manual of BBL® Products and Laboratory Procedures, 6th Ed.* Becton Dickinson Microbiology Systems, Cockeysville, MD.

TABLE OF RESULTS	
RESULT	**INTERPRETATION**
Poor growth or no growth	Organism is inhibited by phenylethyl alcohol
Good growth	Organism is not inhibited by phenylethyl alcohol

ORGANISM	RESULT	REMARKS

SELECTIVE MEDIA FOR ISOLATION OF GRAM-NEGATIVE RODS

The selective media appearing in the following exercises are used for isolation of Gram-negative rods, especially the enterics. While inhibiting growth of Gram-positive cocci, the media also differentiate pathogenic and nonpathogenic Gram-negative rods likely to be found together in feces or wound infections. The media in these exercises is a small sampling of those actually used in the clinical laboratory (see Section 2 of the *Photographic Atlas*). Although many media perform similar functions, no two are exactly alike. Each has advantages and disadvantages which make it suitable for some applications and unsuitable for others.

Desoxycholate agar, eosin methylene blue agar and MacConkey agar are used to differentiate lactose-fermenting Gram-negative bacilli from lactose non-fermenters. Hektoen enteric agar and XLD are used to differentiate *Salmonella* and *Shigella* from other enterics based on their ability to overcome inhibitory factors, to ferment carbohydrates, and to produce H_2S. Read the purposes and principles of the tests in the *Photographic Atlas* and compare the effectiveness of each medium to the others as you perform the exercises.

Exercise 4–5   Desoxycholate (DOC) Agar

MATERIALS

One DOC plate
Recommended organisms:
  Staphylococcus aureus
  Klebsiella pneumoniae
  Escherichia coli

TEST PROTOCOL

1. Use a marking pen to divide the plate into four sectors. Marks should be made on the plate's base.

2. Spot inoculate in different sectors and leave the fourth sector uninoculated as a control.

3. Label the plate with the organisms' names, your name, and the date.

4. Invert the plate and incubate at 35°C for 24 to 48 hours.

5. Examine the plate and, using the information in the Table of Results, record your results in the space provided (see the *Photographic Atlas*, page 10).

> **Photographic Atlas Reference Page 10**

PRECAUTION

⚠ The medium is heat sensitive. It should not be autoclaved or overheated when prepared, nor can it be remelted and repoured.

REFERENCES

DIFCO Laboratories. 1984. Page 274 in *DIFCO Manual, 10th Ed.* DIFCO Laboratories, Detroit, MI.

Koneman, Elmer W., Stephen D. Allen, William M. Janda, Paul C. Schreckenberger and Washington C. Winn, Jr. 1997. Chapters 4 and 6 in *Color Atlas and Textbook of Diagnostic Microbiology, 5th Ed.* J.B. Lippincott Company, Philadelphia, PA.

Power, David A. and Peggy J. McCuen. 1988. Page 144 in *Manual of BBL® Products and Laboratory Procedures, 6th Ed.* Becton Dickinson Microbiology Systems, Cockeysville, MD.

TABLE OF RESULTS	
RESULT	**INTERPRETATION**
Poor growth or no growth	Organism is inhibited by desoxycholate
Good growth	Organism is not inhibited by desoxycholate
Growth is pink	Organism ferments lactose to acid end products
Growth is colorless	Organism does not ferment lactose

ORGANISM	RESULT	REMARKS

Exercise 4-6

Eosin Methylene Blue (EMB) Agar

MATERIALS

One EMB plate
Recommended organisms:
    Escherichia coli
    Klebsiella pneumoniae
    Proteus vulgaris
    Staphylococcus aureus

TEST PROTOCOL

1. Use a marking pen to divide the plate into four sectors. Marks should be made on the plate's base.

2. Spot inoculate in different sectors. (Spot inoculations are for demonstration only. This medium is intended to be streaked for isolation or used in the Membrane Filter Technique — Exercise 8-4.)

3. Label the plate with the organisms' names, your name, and the date.

4. Invert the plate and incubate at 35°C for 24 to 48 hours.

5. Examine the plate and, using the information in the Table of Results, record your results in the space provided (see the *Photographic Atlas*, page 12).

PRECAUTION

⚠ Negative results after 24 hours should be incubated for another 24 hours.

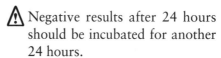
Photographic Atlas Reference Page 12

REFERENCES

DIFCO Laboratories. 1984. Page 324 in *DIFCO Manual, 10th Ed.* DIFCO Laboratories, Detroit, MI.

Koneman, Elmer W., Stephen D. Allen, William M. Janda, Paul C. Schreckenberger and Washington C. Winn, Jr. 1997. Chapter 4 in *Color Atlas and Textbook of Diagnostic Microbiology, 5th Ed.* J.B. Lippincott Company, Philadelphia, PA.

Power, David A. and Peggy J. McCuen. 1988. Page 153 in *Manual of BBL® Products and Laboratory Procedures, 6th Ed.* Becton Dickinson Microbiology Systems, Cockeysville, MD.

TABLE OF RESULTS	
RESULT	**INTERPRETATION**
Poor growth or no growth	Organism is inhibited by eosin and methylene blue
Good growth	Organism is not inhibited by eosin and methylene blue
Growth is pink and mucoid	Organism ferments lactose with little acid production
Growth is purple to black, with or without green sheen	Organism ferments lactose with much acid production
Growth is colorless	Organism does not ferment lactose or sucrose

ORGANISM	RESULT	REMARKS

Hektoen Enteric (HE) Agar

MATERIALS

One Hektoen enteric agar plate
Recommended organisms:
  Escherichia coli
  Proteus vulgaris
  Salmonella typhimurium

TEST PROTOCOL

1. Use a marking pen to divide the plate into four sectors. Marks should be made on the plate's base.

2. Spot inoculate in different sectors and leave the fourth sector uninoculated as a control.

3. Label the plate with the organisms' names, your name, and the date.

4. Invert the plate and incubate at 35°C for 24 to 48 hours away from the light.

5. Examine the plate and, using the information in the Table of Results, record your results below.

PRECAUTIONS

⚠ The agar surface should be dry prior to inoculation.

⚠ Negative results after 24 hours should be incubated for another 24 hours.

Photographic Atlas
Reference
Page 13

REFERENCES

Baron, Ellen Jo, Lance R. Peterson and Sydney M. Finegold. 1994. Chapter 9 in *Bailey and Scott's Diagnostic Microbiology, 9th Ed.* Mosby-Yearbook, St. Louis, MO.

DIFCO Laboratories. 1984. Page 455 in *DIFCO Manual, 10th Ed.* DIFCO Laboratories, Detroit, MI.

Koneman, Elmer W., Stephen D. Allen, William M. Janda, Paul C. Schreckenberger and Washington C. Winn, Jr. 1997. Chapter 4 in *Color Atlas and Textbook of Diagnostic Microbiology, 5th Ed.* J.B. Lippincott Company, Philadelphia, PA.

Power, David A. and Peggy J. McCuen. 1988. Page 167 in *Manual of BBL® Products and Laboratory Procedures, 6th Ed.* Becton Dickinson Microbiology Systems, Cockeysville, MD.

TABLE OF RESULTS	
RESULT	**INTERPRETATION**
Poor growth or no growth	Organism is inhibited by bile and/or one of the dyes included
Good growth	Organism is not inhibited by bile or any of the dyes included
Green or blue-green growth	Organism does not ferment lactose
Pink to orange growth	Organism produces acid from lactose fermentation
Growth has black center	Organism produces H_2S from sulfur reduction

ORGANISM	RESULT	REMARKS

Exercise 4–8

MacConkey Agar

MATERIALS

One MacConkey agar plate
Recommended organisms:
    Enterobacter aerogenes
    Enterococcus faecalis
    Proteus vulgaris

TEST PROTOCOL

1. Use a marking pen to divide the plate into four sectors. Marks should be made on the plate's base.

2. Spot inoculate in different sectors and leave the fourth sector uninoculated as a control.

3. Label the plate with the organisms' names, your name, and the date.

4. Invert the plate and incubate at 35°C for 24 to 48 hours.

5. Examine the plate and, using the information in the Table of Results, record your results in the space provided.

PRECAUTIONS

⚠ The agar surface should be dry prior to inoculation.

⚠ Negative results after 24 hours should be incubated for another 24 hours.

Photographic Atlas Reference Page 14

REFERENCES

Baron, Ellen Jo, Lance R. Peterson and Sydney M. Finegold. 1994. Chapters 9 and 28 in *Bailey and Scott's Diagnostic Microbiology, 9th Ed.* Mosby-Yearbook, St. Louis, MO.

DIFCO Laboratories. 1984. Page 546 in *DIFCO Manual, 10th Ed.* DIFCO Laboratories, Detroit, MI.

Koneman, Elmer W., Stephen D. Allen, William M. Janda, Paul C. Schreckenberger and Washington C. Winn, Jr. 1997. Chapter 4 in *Color Atlas and Textbook of Diagnostic Microbiology, 5th Ed.* J.B. Lippincott Company, Philadelphia, PA.

Power, David A. and Peggy J. McCuen. 1988. Page 189 in *Manual of BBL® Products and Laboratory Procedures, 6th Ed.* Becton Dickinson Microbiology Systems, Cockeysville, MD.

TABLE OF RESULTS	
RESULT	**INTERPRETATION**
Poor growth or no growth	Organism is inhibited by crystal violet and/or bile
Good growth	Organism is not inhibited by crystal violet or bile
Green or red growth	Organism produces acid from lactose fermentation
Colorless growth	Organism does not ferment lactose

ORGANISM	RESULT	REMARKS

Exercise 4–9

Xylose Lysine Desoxycholate (XLD) Agar

MATERIALS

One XLD plate
Recommended organisms:
  Providencia stuartii
  Salmonella typhimurium
  Enterobacter aerogenes

TEST PROTOCOL

1. Use a marking pen to divide the plate into four sectors. Marks should be made on the plate's base.

2. Spot inoculate in different sectors and leave the fourth sector uninoculated as a control.

3. Label the plate with the organisms' names, your name, and the date.

4. Invert the plate and incubate at 35°C for 18 to 24 hours.

5. Examine the plate and, using the information in the Table of Results, record your results in the space provided.

PRECAUTIONS

⚠ Avoid overheating the medium while preparing it and do not autoclave it. Heat causes a precipitate to form.

⚠ Incubation periods longer than 24 hours may lead to false identification of *Shigella* species, which are distinguished from other enterics by their inability to ferment xylose — and so produce a red color on XLD. Some microbes ferment the carbohydrates in XLD to acid and lower the pH, then decarboxylate the lysine with a subsequent raising of the pH which produces a red color.

Photographic Atlas Reference Page 20

REFERENCES

Baron, Ellen Jo, Lance R. Peterson and Sydney M. Finegold. 1994. Chapter 9 in *Bailey and Scott's Diagnostic Microbiology, 9th Ed.* Mosby-Yearbook, St. Louis, MO.

DIFCO Laboratories. 1984. Page 1128 in *DIFCO Manual, 10th Ed.* DIFCO Laboratories, Detroit, MI.

Power, David A. and Peggy J. McCuen. 1988. Page 288 in *Manual of BBL® Products and Laboratory Procedures, 6th Ed.* Becton Dickinson Microbiology Systems, Cockeysville, MD.

TABLE OF RESULTS	
RESULT	**INTERPRETATION**
Poor growth	Organism is inhibited by desoxycholate
Good growth	Organism is not inhibited by desoxycholate
Growth is red	Organism does not ferment xylose or ferments xylose slowly Organism produces alkaline products from lysine decarboxylation
Growth is yellow	Organism produces acid from xylose fermentation
Growth has black center	Organism produces H_2S from sulfur reduction

ORGANISM	RESULT	REMARKS

Differential Tests

Bacteria appear in three basic morphologies — rod shaped, spherical, and helical or spiral. There are wide variations on these basic shapes but the subtle differences are usually not sufficient to allow more than presumptive identification. On the other hand, bacterial biochemistry demonstrates much greater diversity. It is this diversity of which *differential media* takes advantage.

Differential media are growth media formulated to detect certain biochemical characteristics. If the medium contains ingredients which favor growth of one group of organisms while inhibiting another it is called differential and *selective*. Technological advances have introduced many new and rapid methods of bacterial identification, however, many of the multiple tests and rapid enzyme identification systems still rely on the time-tested principles that are the mainstay of traditional biochemical tests.

The tests, media and bacteria included in this section were chosen to demonstrate major metabolic pathways. You will be performing tests which demonstrate the following:

1. Energy metabolism including fermentation of carbohydrates *and* aerobic and anaerobic respiration

2. Utilization of a specified medium component

3. Decarboxylation and deamination of amino acids

4. Hydrolytic reactions requiring intracellular and extracellular enzymes

5. Multiple reactions performed by a single combination medium

6. Antimicrobial susceptibility

7. Miscellaneous differential tests

In an attempt to enable you to produce results not wholly predictable we have (when possible) recommended bacteria which do not appear in corresponding *Photographic Atlas* exercises. In some cases, however, we thought it more fitting to recommend the typical positive and negative control organisms.

INTRODUCTION TO ENERGY METABOLISM TESTS

Bacteria produce ATP by means of three major metabolic pathways: photosynthesis, respiration (both aerobic and anaerobic), and fermentation. Since photoautotrophic bacteria are generally not studied in an introductory microbiology laboratory, we will consider only respiration and fermentation performed by heterotrophs.

In this first unit, you will perform a simple test used to determine the ability of an organism to perform oxidative and/or fermentative metabolism of sugars. In the exercises which follow, you will perform tests which demonstrate microbial fermentative characteristics and tests designed to detect specific respiratory constituents or pathways.

Exercise 5–1

Oxidation-Fermentation (OF) Medium

MATERIALS

Eight OF glucose tubes
Eight OF lactose tubes
Eight OF sucrose tubes
Sterile mineral oil
Sterile Pasteur pipettes
Recommended organisms:
> *Alcaligenes faecalis*
> *Enterobacter aerogenes*
> *Acinetobacter calcoaceticus*

TEST PROTOCOL

1. Stab inoculate two OF tubes (of each carbohydrate broth) with each organism being tested. Using a light inoculum, stab to a depth of about 0.5 cm from the bottom of the agar. Leave two tubes of each carbohydrate broth uninoculated as controls.

2. Overlay one of each pair of tubes (including the controls) with 3–4 mm of sterile mineral oil (Figure 5-1).

3. Label the tubes with the organisms' names, your name and the date.

4. Incubate the tubes with both uninoculated controls at 35°C for 48 hours.

5. Examine the tubes for characteristic yellow color formation. Reincubate any OF tubes that have not changed color for an additional 48 hours.

> *Photographic Atlas Reference Page 73*

6. Record your results in the space provided on the following page.

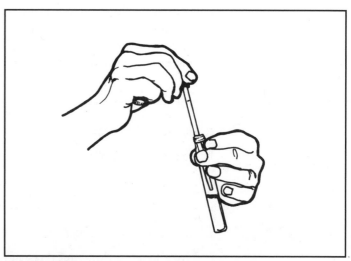

FIGURE 5-1 **Adding the Mineral Oil Layer**
Tip the tube slightly to one side and gently add 2 to 3 mL of sterile mineral oil. Be sure to use a sterile pipette for *each* tube.

TABLE OF RESULTS			
RESULT		INTERPRETATION	SYMBOL
SEALED	UNSEALED		
Green or blue	Green or blue	No sugar metabolism ("nonsaccharolytic")	N
Green or blue	Yellow	Oxidative metabolism	O
Yellow	Yellow	Oxidative and/or fermentative metabolism	O/F

ORGANISM	GLUCOSE		LACTOSE		SUCROSE	
	SEALED	UNSEALED	SEALED	UNSEALED	SEALED	UNSEALED

REFERENCES

Collins, C. H., Patricia M. Lyne, J. M. Grange. 1995. Page 112 in *Collins and Lyne's Microbiological Methods, 7th Ed.* Butterworth-Heinemann, UK.

DIFCO Laboratories. 1984. Page 625 in *DIFCO Manual, 10th Ed.* DIFCO Laboratories, Detroit, MI.

MacFaddin, Jean F. 1980. Page 260 in *Biochemical Tests for Identification of Medical Bacteria, 2nd Ed.* Williams & Wilkins, Baltimore, MD.

Power, David A. and Peggy J. McCuen. 1988. Page 216 in *Manual of BBL® Products and Laboratory Procedures, 6th Ed.* Becton Dickinson Microbiology Systems, Cockeysville, MD.

Smibert, Robert M. and Noel R. Krieg. 1994. Page 625 in *Methods for General and Molecular Bacteriology*, edited by Philipp Gerhardt, R. G. E. Murray, Willis A. Wood and Noel R. Krieg, American Society for Microbiology, Washington, DC.

TESTS IDENTIFYING MICROBIAL ABILITY TO FERMENT

Fermentation (in contrast to respiration) is the complex metabolic process by which an organic molecule (*i.e.*, glucose) acts as an electron donor and one or more of its organic products act as the final electron acceptor. Reduced carbon compounds in the form of acids, organic solvents, and sometimes CO_2 are the typical end products of fermentation.

Differential media formulated to detect fermentation typically include one or more fermentable carbohydrates and a pH indicator to detect acid formation. Purple broth and phenol red broth are examples of general-purpose fermentation media. MR/VP broth is a more specialized medium. The methyl red test detects bacteria capable of forming multiple acids ("mixed acids"), possible only by the process of fermentation. The Voges-Proskauer test identifies the ability of a microbe to produce acetoin as part of a 2,3-butanediol fermentation. Methyl red and Voges-Proskauer are typically run simultaneously in MR/VP broth.

Exercise 5–2 — Fermentation Tests (Phenol Red Broth)

MATERIALS

Four PR glucose broths with Durham tubes
Four PR lactose broths with Durham tubes
Four PR sucrose broths with Durham tubes
18 to 24 hour pure cultures of:
  Klebsiella pneumoniae
  Pseudomonas aeruginosa
  Streptococcus faecalis

TEST PROTOCOL

1. Inoculate broths of each carbohydrate with the test organisms. Leave one tube of each uninoculated as a control. (See Precautions.)

2. Label the tubes with the organisms' names, your name and the date.

Photographic Atlas Reference Page 52

3. Incubate all the tubes (including the uninoculated control of each carbohydrate broth) aerobically at 35°C for 24 to 48 hours.

4. Examine the tubes for acid and/or gas production or alkaline reactions.

5. Record your results in the table below.

PRECAUTIONS

⚠ Avoid overheating this medium and remove it from the autoclave immediately after sterilization.

TABLE OF RESULTS		
RESULT	**INTERPRETATION**	**SYMBOL**
Yellow broth, bubble in tube	Fermentation with acid and gas end products	A/G
Yellow broth, no bubble in tube	Fermentation with acid, but no gas end products	A/–
Red broth, no bubble in tube	No fermentation	–/–
Fuchsia broth, no bubble in tube	Degradation of peptone (either in place of or following fermentation)	K

ORGANISM	PR GLUCOSE	PR LACTOSE	PR SUCROSE

⚠ Don't forget to compare your results with the uninoculated control. This test is sometimes difficult to read, especially with weak alkaline reactions. The best results are obtained when inoculating with fresh bacterial cultures.

⚠ Some actively fermenting bacteria will cause a reversion from acidity to alkalinity of this medium after 48 hours. If this occurs, either repeat the test and read it after 24 hours or increase the carbohydrate to 1%.

REFERENCES

DIFCO Laboratories. 1984. Page 660 in *DIFCO Manual, 10th Ed.* DIFCO Laboratories, Detroit, MI.

Lányi, B. 1987. Page 44 in *Methods in Microbiology, Vol. 19*, edited by R. R. Colwell and R. Grigorova, Academic Press Inc., New York.

MacFaddin, Jean F. 1980. Page 36 in *Biochemical Tests for Identification of Medical Bacteria, 2nd Ed.* Williams & Wilkins, Baltimore, MD.

Power, David A. and Peggy J. McCuen. 1988. Page 220 in *Manual of BBL® Products and Laboratory Procedures, 6th Ed.* Becton Dickinson Microbiology Systems, Cockeysville, MD.

Exercise 5–3 — Fermentation Tests (Purple Broth)

MATERIALS

Four purple glucose broths with Durham tubes
Four purple lactose broths with Durham tubes
Four purple sucrose broths with Durham tubes
Recommended organisms:
  Escherichia coli
  Pseudomonas aeruginosa
  Staphylococcus aureus

TEST PROTOCOL

1. Inoculate broths of each carbohydrate with the test organisms. Leave one tube of each uninoculated as a control. (See Precautions.)

2. Label the tubes with the organisms' names, your name and the date.

3. Incubate all the tubes (including the uninoculated control of each carbohydrate broth) aerobically at 35°C for 24 to 48 hours.

4. Examine the tubes for acid and/or gas production or alkaline reactions.

5. Record your results in the space provided.

PRECAUTIONS

 Avoid overheating this medium and remove it from the autoclave immediately after sterilization.

⚠ Don't forget to compare your results with the uninoculated control. This test is sometimes difficult to read, especially with weak alkaline reactions. The best results are obtained when inoculating with fresh bacterial cultures.

⚠ Some actively fermenting bacteria may cause a reversion from acidity to alkalinity of this medium after 48 hours. If this occurs, the test should be repeated and read at 24 hours.

REFERENCES

DIFCO Laboratories. 1984. Page 712 in *DIFCO Manual, 10th Ed.* DIFCO Laboratories, Detroit, MI.

Power, David A. and Peggy J. McCuen. 1988. Page 229 in *Manual of BBL® Products and Laboratory Procedures, 6th Ed.* Becton Dickinson Microbiology Systems, Cockeysville, MD.

Photographic Atlas Reference Page 52

TABLE OF RESULTS

RESULT	INTERPRETATION	SYMBOL
Yellow broth, bubble in tube	Fermentation with acid and gas end products	A/G
Yellow broth, no bubble in tube	Fermentation with acid, but no gas, end products	A/—
Purple broth, no bubble in tube	No fermentation	—/—
Deeper purple broth than control, no bubble in tube	Degradation of peptone (either in place of or following fermentation	K

ORGANISM	PURPLE GLUCOSE BROTH	PURPLE LACTOSE BROTH	PURPLE SUCROSE BROTH

Methyl Red and Voges-Proskauer (MR/VP) Tests

Photographic Atlas
Reference
Pages 63 and 82

MATERIALS

Three MR/VP broths
Methyl red
VP reagents A and B (Appendix B)
Six nonsterile test tubes
Nonsterile 1 mL pipettes
Recommended organisms:
   Citrobacter diversus
   Serratia marcescens

TEST PROTOCOL (Figure 5-2)

Day One

1. Inoculate two broths with the test cultures and leave one uninoculated as a control.

2. Label the broths with the organisms' names, your name and the date.

3. Incubate the tubes aerobically with the uninoculated control at 35°C for 5 days.

Day Two

1. After incubation, aseptically remove two 1.0 mL aliquots from each broth and place into separate tubes.

2. For each pair of tubes, add the reagents as follows:

Tube #1

  a. Add several drops of methyl red reagent.

  b. Observe for red color formation.

  c. Record and interpret your results using the following tables.

Tube #2
(See Precautions.)

  a. Add 0.6 mL of VP reagent A. Mix well.

  b. Add 0.2 mL of VP reagent B. Mix well.

  c. Observe for red color formation within 5 minutes.

  d. Using the Tables of Results, record your results in the space provided.

TABLE OF METHYL RED RESULTS		
RESULT	**INTERPRETATION**	**SYMBOL**
Red	Mixed acid fermentation	+
No color change	No mixed acid fermentation	−

TABLE OF VOGES-PROSKAUER RESULTS		
RESULT	**INTERPRETATION**	**SYMBOL**
Red	Acetoin produced (2,3-butanediol fermentation)	+
No color change	Acetoin is not produced	−

PRECAUTIONS

⚠ It is essential that the VP reagents be added in the correct order and that none of the measurements be exceeded. False Voges-Proskauer (VP) positives will likely occur if the KOH is added first.

⚠ A common problem with the VP test is differentiating weak positive reactions from negative reactions. VP (−) reactions often produce a copper color; weak VP (+) reactions produce a pink color. The strongest color change should be at the surface.

REFERENCES

DIFCO Laboratories. 1984. Page 543 in *DIFCO Manual, 10th Ed.* DIFCO Laboratories, Detroit, MI.

MacFaddin, Jean F. 1980. Pages 209 and 308 in *Biochemical Tests for Identification of Medical Bacteria, 2nd Ed.* Williams & Wilkins, Baltimore, MD.

Power, David A. and Peggy J. McCuen. 1988. Page 202 in *Manual of BBL® Products and Laboratory Procedures, 6th Ed.* Becton Dickinson Microbiology Systems, Cockeysville, MD.

Smibert, Robert M. and Noel R. Krieg. 1994. Pages 622 and 630 in *Methods for General and Molecular Bacteriology*, edited by Philipp Gerhardt, R. G. E. Murray, Willis A. Wood and Noel R. Krieg, American Society for Microbiology, Washington, DC.

ORGANISM	MR +/−	VP +/−

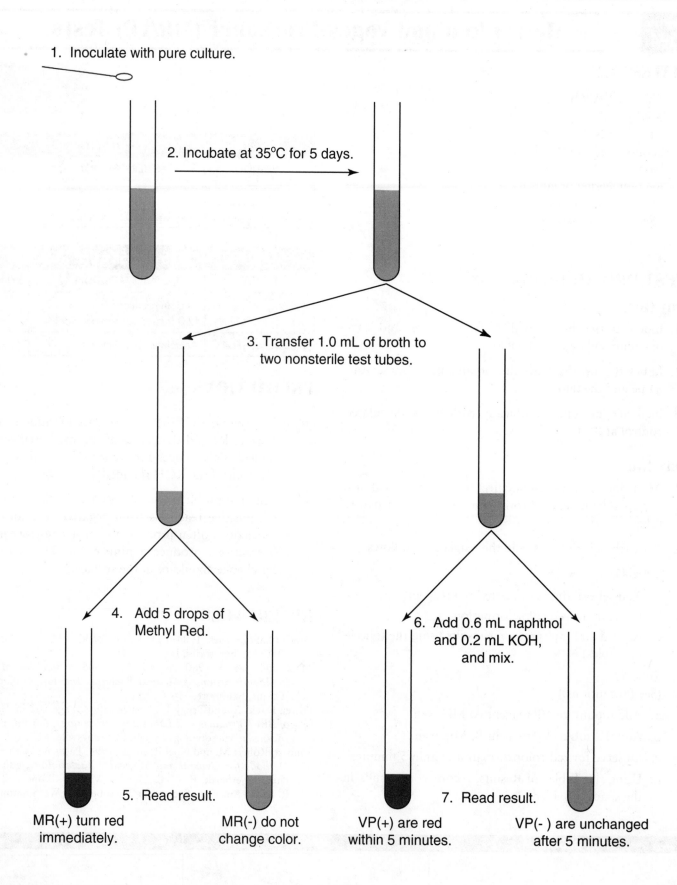

1. Inoculate with pure culture.

2. Incubate at 35°C for 5 days.

3. Transfer 1.0 mL of broth to two nonsterile test tubes.

4. Add 5 drops of Methyl Red.

5. Read result.

MR(+) turn red immediately.

MR(-) do not change color.

6. Add 0.6 mL naphthol and 0.2 mL KOH, and mix.

7. Read result.

VP(+) are red within 5 minutes.

VP(-) are unchanged after 5 minutes.

FIGURE 5-2 **Procedural Diagram for the Methyl Red and Voges-Proskauer Tests**

TESTS IDENTIFYING MICROBIAL ABILITY TO RESPIRE

Respiration is the complex catabolic process by which glucose is converted to CO_2, H_2O (or some other reduced inorganic chemical), and ATP. The three major phases of respiration are glycolysis, the tricarboxylic acid (TCA) cycle (also called the Krebs cycle), and oxidative phosphorylation (also called the electron transport chain). When inorganic oxygen is the final electron acceptor (*i.e.*, reduced) in oxidative phosphorylation, the respiration is said to be *aerobic*. If an inorganic molecule other than oxygen such as sulfate or nitrate is reduced, the respiration is *anaerobic*. Metabolic pathways are described in detail in Appendices A, B, and C of the *Photographic Atlas*.

Many techniques have been devised to differentiate bacteria based on their abilities to respire aerobically or anaerobically. The catalase test detects the ability to produce catalase, an enzyme that detoxifies hydrogen peroxide produced in the aerobic electron transport chain. The oxidase test identifies the presence of cytochrome oxidase in the aerobic electron transport chain. The nitrate reduction test examines bacterial ability to reduce nitrate and thus respire anaerobically. Other media, such as TSI, KIA and SIM, detect sulfur reduction to H_2S as a result of anaerobic respiration. However, since these media can detect non-respiratory H_2S producing pathways, we chose to include them in "Combination Media" also in this section.

Exercise 5–5 — Catalase Test

MATERIALS

Three nutrient agar slants
Hydrogen peroxide (3%)
Droppers
Microscope slides
Recommended organisms:
　Staphylococcus epidermidis
　Streptococcus faecalis

TEST PROTOCOL

1. Streak inoculate two slants with the test organisms and leave the third slant uninoculated as a control.

2. Label the slants with the organisms' names, your name and the date.

3. Incubate the slants aerobically with the uninoculated control at 35°C for 24 hours.

4. After incubation, perform the tests as follows:

   Slide Test

   a. Transfer a large amount of growth to a microscope slide.

   b. Apply 1 or 2 drops of hydrogen peroxide directly to the bacteria on the slide.

   c. Observe for the formation of bubbles.

   Slant Test

   a. Place several drops of hydrogen peroxide directly into the slant containing the fresh culture.

   b. Observe for the formation of bubbles.

5. Record your results in the space provided.

Photographic Atlas Reference Page 44

TABLE OF RESULTS		
RESULT	**INTERPRETATION**	**SYMBOL**
Bubbles	Catalase is present	+
No bubbles	Catalase is absent	−

ORGANISM	RESULT +/−

PRECAUTIONS

⚠ When performing the slide test, place the bacteria on the slide first and then add the hydrogen peroxide. Placing the inoculating needle in the hydrogen peroxide may catalyze a false positive reaction.

⚠ Do not use blood agar as the growth medium; catalase in the blood will result in false positive reactions.

⚠ To reduce the chance of false negative results, use only fresh cultures for this test (24 hours or less).

REFERENCES

Collins, C. H., Patricia M. Lyne, J. M. Grange. 1995. Page 110 in *Collins and Lyne's Microbiological Methods, 7th Ed.,* Butterworth-Heinemann, UK.

DIFCO Laboratories. 1984. Page 619 in *DIFCO Manual, 10th Ed.* DIFCO Laboratories, Detroit, MI.

Lányi, B. 1987. Page 20 in *Methods in Microbiology, Vol. 19,* edited by R. R. Colwell and R. Grigorova, Academic Press Inc., New York.

MacFaddin, Jean F. 1980. Page 51 in *Biochemical Tests for Identification of Medical Bacteria, 2nd Ed.,* Williams & Wilkins, Baltimore, MD.

Smibert, Robert M. and Noel R. Krieg. 1994. Page 614 in *Methods for General and Molecular Bacteriology,* edited by Philipp Gerhardt, R. G. E. Murray, Willis A. Wood and Noel R. Krieg, American Society for Microbiology, Washington, DC.

<table>
<tr><td>**Exercise 5–6**</td><td># Oxidase Test</td></tr>
</table>

MATERIALS

One nutrient agar plate
Test reagent
Filter paper in a Petri dish
Sterile *nonmetal* applicator or loop
Pasteur pipettes
Recommended organisms:
    Escherichia coli
    Moraxella catarrhalis

TEST PROTOCOL

Day One

1. Using a marking pen, divide the nutrient agar plate into three equal sectors. Be sure to mark the bottom of the plate.

2. Spot inoculate two sectors with the test organisms and leave the third sector as a control.

3. Label the plate with the organisms' names, your name and the date.

4. Invert the plate and incubate it at 35°C for 24 hours.

Day Two

Indirect Method (See Precautions.)

1. Place the filter paper in the sterile Petri dish and saturate it with the test reagent.

2. Using the sterile applicator transfer a large amount of fresh growth from the plate to the filter paper.

3. Observe for color change within 10 to 15 seconds. (See the Table of Results below.)

4. Record your results in the space provided.

Direct Method

1. Place a few drops of oxidase test reagent onto the bacterial growth and onto the uninoculated control sector.

2. Observe for color change within 10 to 15 seconds.

3. Record your results in the space provided.

Photographic Atlas Reference Page 71

PRECAUTIONS

⚠ Tetramethyl-*p*-phenylenediamine in solution is very unstable and will auto-oxidize after a short time. Therefore, color changes after 60 seconds are not considered positive.

⚠ The test reagent should be made immediately prior to its use. If it must be stored, keep it in the refrigerator or freezer and allow it to warm to room temperature immediately before use.

REFERENCES

Collins, C. H., Patricia M. Lyne, J. M. Grange. 1995. Page 116 in *Collins and Lyne's Microbiological Methods, 7th Ed.* Butterworth-Heinemann, UK.

Lányi, B. 1987. Page 18 in *Methods in Microbiology, Vol. 19,* edited by R. R. Colwell and R. Grigorova, Academic Press Inc., New York.

MacFaddin, Jean F. 1980. Page 249 in *Biochemical Tests for Identification of Medical Bacteria, 2nd Ed.* Williams & Wilkins, Baltimore, MD.

Smibert, Robert M. and Noel R. Krieg. 1994. Page 625 in *Methods for General and Molecular Bacteriology,* edited by Philipp Gerhardt, R. G. E. Murray, Willis A. Wood and Noel R. Krieg, American Society for Microbiology, Washington, DC.

TABLE OF RESULTS		
RESULT	**INTERPRETATION**	**SYMBOL**
Dark blue	Cytochrome oxidase is present	+
No color change	Cytochrome oxidase is absent	–

ORGANISM	DIRECT +/–	INDIRECT +/–

Nitrate Reduction Test

MATERIALS

Four nitrate reduction broths
Nitrate test reagents A and B (Appendix B)
Zinc powder
Recommended organisms:
Pseudomonas stutzeri
Acinetobacter calcoaceticus
Escherichia coli

TEST PROTOCOL (Figure 5-3)

Day One

1. Inoculate three broths with the test organisms and leave one uninoculated as a control.

2. Label the tubes with the organisms' names, your name and the date.

3. Incubate the tubes aerobically with the uninoculated control at 35°C for 24 to 48 hours.

Day Two

1. After incubation, add the reagents as follows:

   a. Add 0.5 mL of Reagent A and 0.5 mL of reagent B to each tube. Mix well.

   b. Formation of a red color within approximately five minutes indicates reduction of nitrate to nitrite. Record as a (+1).

   c. If there is no color change, add a pinch of zinc dust. Formation of a red color within approximately ten minutes indicates the presence of nitrate, so the test is recorded as negative for nitrate reduction (see Figure 5-60 in the *Photographic Atlas*).

   d. No color change after zinc addition indicates reduction of nitrate beyond nitrite (see Figure 5-61 in the *Photographic Atlas*), and is considered a positive test. Record as a (+2).

Photographic Atlas Reference Page 65

5. Record your results in the space provided.

PRECAUTION

⚠ The most common error is in misinterpreting the results. What does red mean *this* time?

REFERENCES

DIFCO Laboratories. 1984. Page 1023 in *DIFCO Manual, 10th Ed.* DIFCO Laboratories, Detroit, MI.

Lányi, B. 1987. Page 21 in *Methods in Microbiology, Vol. 19*, edited by R. R. Colwell and R. Grigorova, Academic Press Inc., New York.

MacFaddin, Jean F. 1980. Page 236 in *Biochemical Tests for Identification of Medical Bacteria, 2nd Ed.* Williams & Wilkins, Baltimore, MD.

Power, David A. and Peggy J. McCuen. 1988. Page 213 in *Manual of BBL® Products and Laboratory Procedures, 6th Ed.* Becton Dickinson Microbiology Systems, Cockeysville, MD.

TABLE OF RESULTS		
RESULT	**INTERPRETATION**	**SYMBOL**
Red (after addition of Reagents A and B)	Reduction of nitrate to nitrite; denitrification	+1
No color change (after addition of zinc)	Reduction of nitrate to something other than nitrite.	+2
Red (after addition of zinc dust)	Negative for nitrate reduction	–

ORGANISM	RESULT

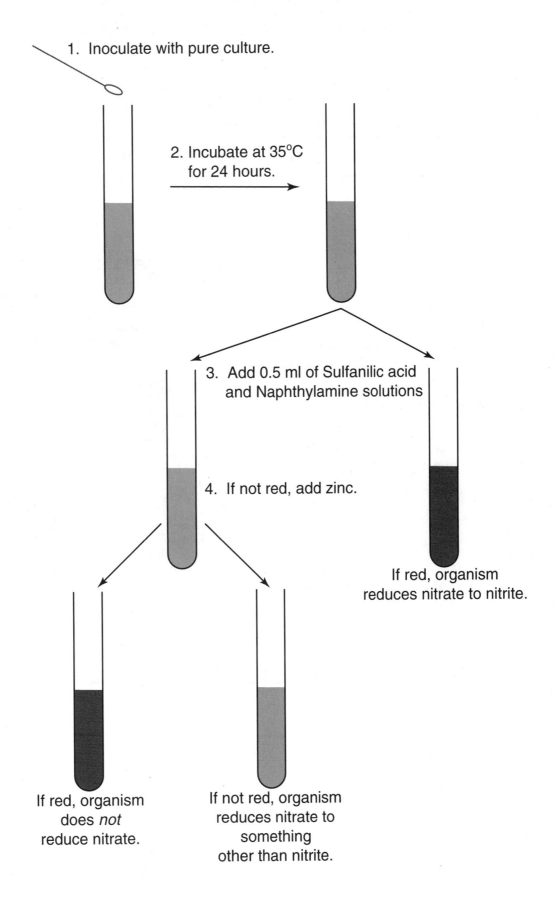

1. Inoculate with pure culture.

2. Incubate at 35°C for 24 hours.

3. Add 0.5 ml of Sulfanilic acid and Naphthylamine solutions

4. If not red, add zinc.

If red, organism reduces nitrate to nitrite.

If red, organism does *not* reduce nitrate.

If not red, organism reduces nitrate to something other than nitrite.

FIGURE 5-3   **Procedural Diagram for the Nitrate Reduction Test**

UTILIZATION MEDIA

Utilization media are carefully defined formulations designed to differentiate organisms based on their abilities to metabolize a specific component of the medium (*e.g.*, carbohydrates, organic acids and amino acids), often in the presence of inhibiting factors. Simmons citrate agar is a utilization medium that detects bacterial ability to use citrate as the sole source of carbon in the presence of inorganic ammonium salts.

Exercise 5–8

Citrate Utilization Test

MATERIALS

Photographic Atlas Reference Page 45

Three Simmons citrate agar slants
Recommended organisms:
    Enterobacter aerogenes
    Escherichia coli

TEST PROTOCOL

1. Using a *light* inoculum, streak inoculate two slants with the test organisms. Leave the third slant uninoculated as a control.

2. Label the slants with the organisms' names, your name and the date.

3. Incubate the tubes aerobically with the uninoculated control at 35°C for up to 4 days.

4. Observe the tubes for color changes.

5. Record your results in the space provided.

PRECAUTION

⚠ Formation of blue color in any portion of the slant is considered a positive reaction.

TABLE OF RESULTS

RESULT	INTERPRETATION	SYMBOL
Prussian blue	Citrate is utilized	+
No color change	Citrate is not utilized	−

REFERENCES

Collins, C. H., Patricia M. Lyne, J. M. Grange. 1995. Page 111 in *Collins and Lyne's Microbiological Methods, 7th Ed.* Butterworth-Heinemann, UK.

DIFCO Laboratories. 1984. Page 864 in *DIFCO Manual, 10th Ed.* DIFCO Laboratories, Detroit, MI.

Power, David A. and Peggy J. McCuen. 1988. Page 246 in *Manual of BBL® Products and Laboratory Procedures, 6th Ed.* Becton Dickinson Microbiology Systems, Cockeysville, MD.

Smibert, Robert M. and Noel R. Krieg. 1994. Page 614 in *Methods for General and Molecular Bacteriology*, edited by Philipp Gerhardt, R. G. E. Murray, Willis A. Wood and Noel R. Krieg, American Society for Microbiology, Washington, DC.

ORGANISM	RESULT + / −

DECARBOXYLATION AND DEAMINATION TESTS

Decarboxylation is the removal of the carboxyl group (COOH) from an amino acid. Deamination is the removal of the amine group (NH_2) from an amino acid or other molecule. Both reactions produce alkaline end products and can be detected by pH indicators. Decarboxylase medium is typically formulated to detect bacterial ability (especially by members of the *Enterobacteriaceae*) to decarboxylate lysine, arginine, and/or ornithine. Phenylalanine deaminase medium detects bacterial ability to deaminate the amino acid phenylalanine.

| Exercise 5–9 | **Decarboxylase Test** |

MATERIALS

Four lysine decarboxylase tubes
Four ornithine decarboxylase tubes
Four arginine decarboxylase tubes
Sterile mineral oil
Recommended organisms:
    Alcaligenes faecalis
    Enterobacter aerogenes
    Pseudomonas aeruginosa

TEST PROTOCOL

1. Lightly inoculate the media with the test organisms, leaving one tube of each amino acid uninoculated as a control.

2. Overlay all tubes (including the controls) with 3 to 4 mm sterile mineral oil (Figure 5-4).

3. Label the tubes with the organisms' names, your name and the date.

4. Incubate the tubes aerobically with the uninoculated control at 35°C for up to one week. More time may be necessary for weak reactions.

5. Remove the tubes from the incubator and examine the tubes for characteristic color changes.

6. Record your results in the space provided.

Photographic Atlas Reference Page 47

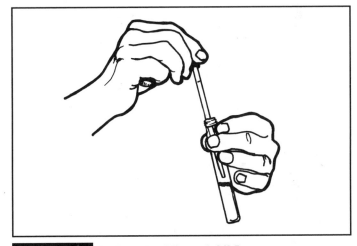

FIGURE 5-4 **Adding the Mineral Oil Layer**
Tip the tube slightly to one side and gently add 2 to 3 mL mineral oil. Be sure to use a sterile pipette for *each* tube.

TABLE OF RESULTS

RESULT	INTERPRETATION	SYMBOL
No color change	Inability to decarboxylate the amino acid in the medium	−
Yellow	Fermentation, but incapable of decarboxylating the amino acid	−
Purple (more than the control)	Decarboxylation (produces the decarboxylase enzyme for the amino acid)	+

ORGANISM	L-LYSINE +/−	L-ORNITHINE +/−	L-ARGININE +/−

PRECAUTIONS

⚠ Do not read this test too early. False negatives may occur if sufficient time has not elapsed.

⚠ Be sure to compare the incubated tubes to an uninoculated control since the color changes may be subtle. A color change to yellow indicates fermentation, *not* decarboxylase activity.

REFERENCES

Collins, C. H., Patricia M. Lyne, J. M. Grange. 1995. Page 111 in *Collins and Lyne's Microbiological Methods, 7th Ed.* Butterworth-Heinemann, UK.

DIFCO Laboratories. 1984. Page 268 in *DIFCO Manual, 10th Ed.* DIFCO Laboratories, Detroit, MI.

Lányi, B. 1987. Page 29 in *Methods in Microbiology, Vol. 19,* edited by R. R. Colwell and R. Grigorova, Academic Press Inc., New York.

MacFaddin, Jean F. 1980. Page 78 in *Biochemical Tests for Identification of Medical Bacteria, 2nd Ed.* Williams & Wilkins, Baltimore, MD.

Exercise 5–10

Phenylalanine Deaminase Test

MATERIALS

Three phenylalanine deaminase agar slants
Ferric chloride test reagent (Appendix B)
Recommended organisms:
    Enterobacter aerogenes
    Proteus vulgaris

TEST PROTOCOL

Day One

1. Streak two slants with the test organisms leaving one slant uninoculated as a control.

2. Label the tubes with the organisms' names, your name and the date.

3. Incubate the slants aerobically with the uninoculated control at 35°C for 18 to 24 hours.

Day Two

1. Add a few drops of test reagent to each tube.

2. Observe for the characteristic green color formation within 1 to 5 minutes.

3. Record your results in the space provided.

PRECAUTION

⚠ The green color resulting from positive reactions in this test will fade quickly, so read and record your results immediately.

Photographic Atlas Reference Page 74

REFERENCES

DIFCO Laboratories. 1984. Page 664 in *DIFCO Manual, 10th Ed.* DIFCO Laboratories, Detroit, MI.

Lányi, B. 1987. Page 28 in *Methods in Microbiology, Vol. 19,* edited by R. R. Colwell and R. Grigorova, Academic Press Inc., New York.

MacFaddin, Jean F. 1980. Page 269 in *Biochemical Tests for Identification of Medical Bacteria, 2nd Ed.* Williams & Wilkins, Baltimore, MD.

Power, David A. and Peggy J. McCuen. 1988. Page 222 in *Manual of BBL® Products and Laboratory Procedures, 6th Ed.* Becton Dickinson Microbiology Systems, Cockeysville, MD.

TABLE OF RESULTS		
RESULT	**INTERPRETATION**	**SYMBOL**
No color change	Phenylalanine deaminase is absent	−
Green color	Phenylalanine deaminase is present	+

ORGANISM	RESULT +/−

TESTS DETECTING THE PRESENCE OF HYDROLYTIC ENZYMES

Hydrolytic enzymes catalyze reactions which split complex molecules. These catabolic reactions require the addition of water to break a covalent bond in the complex molecule — hence the name "hydrolysis." If the reaction takes place inside the cell, the enzyme is considered to be an *intracellular enzyme*. If the enzyme is secreted from the cell and catalyzes reactions in the environment it is an *extracellular enzyme* ("exoenzyme"). In this unit you will be performing tests that identify the actions of both intracellular *and* extracellular enzymes.

The hydrolytic enzymes detected by the urease and bile esculin tests are intracellular and typically identified by color changes in the medium; all other tests in this unit detect the presence of exoenzymes. With the exception of nutrient gelatin, which becomes liquefied by gelatinase-positive organisms, these media are plated to allow observation of a "halo" or clearing around the bacterial colonies. Included in this group are: DNase agar, milk agar, tributyrin agar and spirit blue agar, and starch agar. These media are formulated to detect hydrolysis of DNA, casein, lipids, and starch respectively.

Exercise 5–11	**Bile Esculin Test**

MATERIALS

Three bile esculin agar slants
Recommended organisms:
Proteus mirabilis
Enterococcus faecalis

TEST PROTOCOL

1. Inoculate two slants with the test organisms and leave the third slant uninoculated as a control.

2. Label the slants with the organisms' names, your name and the date.

3. Incubate the slants with the uninoculated control at 35°C for up to 72 hours.

4. After incubation, observe all the tubes for blackening of the agar. Any blackening is scored as positive.

5. Record your results in the space provided.

PRECAUTION

⚠ Incubate for the full 72 hours to reduce the chance of false negatives.

Photographic Atlas Reference Page 40

REFERENCES

DIFCO Laboratories. 1984. Page 129 in *DIFCO Manual, 10th Ed.* DIFCO Laboratories, Detroit, MI.

Lányi, B. 1987. Page 56 in *Methods in Microbiology, Vol. 19*, edited by R. R. Colwell and R. Grigorova, Academic Press Inc., New York.

MacFaddin, Jean F. 1980. Page 4 in *Biochemical Tests for Identification of Medical Bacteria, 2nd Ed.* Williams & Wilkins, Baltimore, MD.

Power, David A. and Peggy J. McCuen. 1988. Page 113 in *Manual of BBL® Products and Laboratory Procedures, 6th Ed.* Becton Dickinson Microbiology Systems, Cockeysville, MD.

TABLE OF RESULTS

RESULT	INTERPRETATION	SYMBOL
Black	Esculin is hydrolyzed in the presence of bile	+
No color change	Esulin is not hydrolyzed	−

ORGANISM	RESULT +/−

Urease Tests (Agar)

MATERIALS

Three urease agar slants
Recommended organisms:
  Enterobacter aerogenes
  Proteus vulgaris

TEST PROTOCOL

1. Streak inoculate two slants with the test organisms, covering the entire agar surface with a heavy inoculum. Do not stab the butt.

2. Label the tubes with the organisms' names, your name and the date.

3. Incubate aerobically with the uninoculated control at 35°C for 6 days.

4. Observe the slants after 2 hours, 6 hours and at 24 hour intervals thereafter.

5. Using the symbols provided, record your observations in the table below.

REFERENCES

Collins, C. H., Patricia M. Lyne, J. M. Grange. 1995. Page 117 in *Collins and Lyne's Microbiological Methods*, 7th Ed. Butterworth-Heinemann, UK.

DIFCO Laboratories. 1984. Page 1040 in *DIFCO Manual*, 10th Ed. DIFCO Laboratories, Detroit, MI.

Lányi, B. 1987. Page 24 in *Methods in Microbiology, Vol. 19*, edited by R. R. Colwell and R. Grigorova, Academic Press Inc., New York.

MacFaddin, Jean F. 1980. Page 298 in *Biochemical Tests for Identification of Medical Bacteria*, 2nd Ed. Williams & Wilkins, Baltimore, MD.

Power, David A. and Peggy J. McCuen. 1988. Page 280 in *Manual of BBL® Products and Laboratory Procedures*, 6th Ed. Becton Dickinson Microbiology Systems, Cockeysville, MD.

Photographic Atlas Reference Page 81

TABLE OF RESULTS

RESULT	INTERPRETATION	SYMBOL
Pink throughout the agar (within 24 hours)	Rapid urea hydrolysis; strong urease production	+4
Pink in slant only (up to 6 days)	Slow urea hydrolysis; weak urease production	+2
Pink in slant tip only (up to 6 days)	Slow urea hydrolysis; weak urease production	+1
No color change (up to 6 days)	Urease is absent	–

ORGANISM	2 HRS	6 HRS	24 HRS	2 DAYS	3 DAYS	4 DAYS	5 DAYS	6 DAYS

Urease Tests (Broth)

Exercise 5–13

MATERIALS

Three urease broths
Recommended organisms:
  Enterobacter aerogenes
  Proteus vulgaris

TEST PROTOCOL

1. Inoculate two broths with a heavy inoculum from the test organisms. Leave the third broth uninoculated as a control.

2. Label the tubes with the organisms' names, your name and the date.

3. Incubate the tubes with the uninoculated control at 35°C for 24 to 48 hours.

4. Observe the tubes for color changes.

5. Record your results in the space provided.

REFERENCES

Collins, C. H., Patricia M. Lyne, J. M. Grange. 1995. Page 117 in *Collins and Lyne's Microbiological Methods, 7th Ed.* Butterworth-Heinemann, UK.

DIFCO Laboratories. 1984. Page 1043 in *DIFCO Manual, 10th Ed.* DIFCO Laboratories, Detroit, MI.

MacFaddin, Jean F. 1980. Page 298 in *Biochemical Tests for Identification of Medical Bacteria, 2nd Ed.* Williams & Wilkins, Baltimore, MD.

Power, David A. and Peggy J. McCuen. 1988. Page 281 in *Manual of BBL® Products and Laboratory Procedures, 6th Ed.* Becton Dickinson Microbiology Systems, Cockeysville, MD.

Smibert, Robert M. and Noel R. Krieg. 1994. Page 630 in *Methods for General and Molecular Bacteriology*, edited by Philipp Gerhardt, R. G. E. Murray, Willis A. Wood and Noel R. Krieg, American Society for Microbiology, Washington, DC.

Photographic Atlas Reference Page 81

TABLE OF RESULTS

RESULT	INTERPRETATION	SYMBOL
No color change	Urease is absent	–
Pink color	Urease is present	+

ORGANISM	RESULT +/–

<table>
<tr><td>Exercise
5-14</td><td></td></tr>
</table>

Casease Test

Materials
One milk agar plate
Recommended organisms:
   Bacillus subtilis
   Escherichia coli

TEST PROTOCOL

1. Using a marking pen, divide the plate into three equal sectors. Be sure to mark on the bottom of the plate.

2. Spot inoculate two sectors with the test organisms and leave the third sector uninoculated as a control.

3. Label the plate with the organisms' names, your name and the date.

4. Invert the plate and incubate it aerobically at 35°C for 24 hours.

5. Examine the plates for clearing around the bacterial growth.

6. Record your results in the space provided.

TABLE OF RESULTS		
RESULT	**INTERPRETATION**	**SYMBOL**
Clearing in agar	Casease is present	+
No clearing in agar	Casease is absent	−

ORGANISM	RESULT +/−

REFERENCES

Chan, E. C. S., Michael J. Pelczar, Jr. and Noel R Krieg. 1986. Page 137 in *Laboratory Exercises In Microbiology*, 5th Ed. McGraw-Hill Book Company.

DIFCO Laboratories. 1984. Page 619 in *DIFCO Manual, 10th Ed.* DIFCO Laboratories, Detroit, MI.

Holt, John G. (Editor). 1994. *Bergey's Manual of Determinative Bacteriology, 9th Ed.* Williams and Wilkins, Baltimore, MD.

Smibert, Robert M. and Noel R. Krieg. 1994. Page 613 in *Methods for General and Molecular Bacteriology*, edited by Philipp Gerhardt, R. G. E. Murray, Willis A. Wood and Noel R. Krieg, American Society for Microbiology, Washington, DC.

Photographic Atlas
Reference
Page 43

DNase Test

MATERIALS

One DNase test agar plate
Recommended organisms:
　Enterobacter aerogenes
　Serratia marcescens

TEST PROTOCOL

1. Using a marking pen, divide the DNase test agar plate into three equal sectors. Be sure to mark the bottom of the plate.

2. Spot inoculate two sectors with the test organisms and leave the third sector as a control.

3. Label the plate with the organisms' names, your name and the date.

4. Invert the plate and incubate it aerobically at 35°C for 24 hours.

5. Examine the plates for clearing around the bacterial growth. (See Precautions.)

6. Record your results in the space provided.

PRECAUTION

 DNase activity usually occurs quickly. Be sure to read this test no more than 24 hours after inoculating to prevent excessive clearing.

**Photographic Atlas
Reference
Page 49**

REFERENCES

Collins, C. H., Patricia M. Lyne, J. M. Grange. 1995. Page 114 in *Collins and Lyne's Microbiological Methods, 7th Ed.* Butterworth-Heinemann, UK.

DIFCO Laboratories. 1984. Page 263 in *DIFCO Manual, 10th Ed.* DIFCO Laboratories, Detroit, MI.

Lányi, B. 1987. Page 33 in *Methods in Microbiology, Vol. 19,* edited by R. R. Colwell and R. Grigorova, Academic Press Inc., New York.

MacFaddin, Jean F. 1980. Page 94 in *Biochemical Tests for Identification of Medical Bacteria, 2nd Ed.* Williams & Wilkins, Baltimore, MD.

Power, David A. and Peggy J. McCuen. 1988. Page 147 in *Manual of BBL® Products and Laboratory Procedures, 6th Ed.* Becton Dickinson Microbiology Systems, Cockeysville, MD.

TABLE OF RESULTS

RESULT	INTERPRETATION	SYMBOL
Clearing in agar	DNase is present	+
No clearing in agar	DNase is absent	−

ORGANISM	RESULT +/−

Gelatin Liquefaction Test

MATERIALS

Three nutrient gelatin stab tubes
Recommended organisms:
   Bacillus subtilis
   Escherichia coli

TEST PROTOCOL

1. Stab inoculate two nutrient gelatin tubes with heavy inocula of test organisms. Leave the third tube uninoculated as a control.

2. Label the tubes with the organisms' names, your name and the date.

3. Incubate the tubes aerobically along with the uninoculated control tube at room temperature for up to one week.

4. Examine the control tube. If the control tube is solid, the test can be read. If the control tube has become liquefied due to the temperature, all tubes must be refrigerated or otherwise cooled until the control is resolidified.

5. Examine the tubes for gelatin liquefaction.

6. Record your results in the space provided.

PRECAUTIONS

⚠ Nutrient gelatin can be incubated at 35°C but it will melt at this temperature. Therefore, incubate the inoculated media in parallel with an uninoculated control and, as described above, cool the media before reading.

⚠ Some organisms take up to 6 weeks to liquefy the gelatin. When incubating for such long time periods, take care to minimize evaporation of the liquefied medium.

> **Photographic Atlas Reference Page 54**

REFERENCES

Collins, C. H., Patricia M. Lyne, J. M. Grange. 1995. Page 112 in *Collins and Lyne's Microbiological Methods, 7th Ed.* Butterworth-Heinemann, UK.

DIFCO Laboratories. 1984. Page 35 in *DIFCO Manual, 10th Ed.* DIFCO Laboratories, Detroit, MI.

Lányi, B. 1987. Page 44 in *Methods in Microbiology, Vol. 19,* edited by R. R. Colwell and R. Grigorova, Academic Press Inc., New York.

MacFaddin, Jean F. 1980. Page 128 in *Biochemical Tests for Identification of Medical Bacteria, 2nd Ed.* Williams & Wilkins, Baltimore, MD.

Power, David A. and Peggy J. McCuen. 1988. Page 215 in *Manual of BBL® Products and Laboratory Procedures, 6th Ed.* Becton Dickinson Microbiology Systems, Cockeysville, MD.

Smibert, Robert M. and Noel R. Krieg. 1994. Page 617 in *Methods for General and Molecular Bacteriology,* edited by Philipp Gerhardt, R. G. E. Murray, Willis A. Wood and Noel R. Krieg, American Society for Microbiology, Washington, DC.

TABLE OF RESULTS		
RESULT	**INTERPRETATION**	**SYMBOL**
Gelatin is liquid (control is solid)	Gelatinase is present	+
Gelatin is solid	Gelatinase is absent	−

ORGANISM	RESULT +/−

MATERIALS

One tributyrin agar plate
One spirit blue agar plate
Recommended organisms:
   Enterobacter aerogenes
   Moraxella catarrhalis
   Staphylococcus aureus
   Proteus mirabilis

TEST PROTOCOL

1. Using a marking pen, divide each plate into three equal sectors. Be sure to mark on the bottoms of the plates.

2. Spot inoculate two sectors of the tributyrin agar plate with *E. aerogenes* and *M. catarrhalis* leaving the third sector uninoculated as a control.

3. Inoculate the spirit blue agar plate in the same manner with *S. aureus* and *P. mirabilis*.

4. Label the plates with the organisms' names, your name and the date.

5. Invert the plates and incubate them aerobically at 35°C for 48 hours.

6. Examine the plates for clearing (or halos) around the bacterial growth.

7. Record your results in the space provided.

PRECAUTION

⚠ The tributyrin oil in the medium will etch the plastic petri dishes after a few days. Since this makes reading results more difficult, we recommend using these plates within 48 hours after preparation.

REFERENCES

Collins, C. H., Patricia M. Lyne, J. M. Grange. 1995. Page 114 in *Collins and Lyne's Microbiological Methods, 7th Ed.* Butterworth-Heinemann, UK.
DIFCO Laboratories. 1984. Page 619 in *DIFCO Manual, 10th Ed.* DIFCO Laboratories, Detroit, MI.
Knapp, Joan S. and Roselyn J. Rice. 1995. Page 335 in *Manual of Cinical. Microbiology, 6th Ed.,* edited by Patrick R. Murray, Ellen Jo Baron, Michael A. Pfaller, Fred C. Tenover and Robert H. Yolken, ASM Press, Washington, DC.

TABLE OF RESULTS		
RESULT	**INTERPRETATION**	**SYMBOL**
Clearing in agar	Lipase is present	+
No clearing in agar	Lipase is absent	−

MEDIUM	ORGANISM	RESULT +/−

**Exercise
5–18**

Starch Hydrolysis Test

MATERIALS

One starch agar plate
Gram iodine (from your Gram stain kit)
Recommended organisms:
   Bacillus subtilis
   Staphylococcus aureus

TEST PROTOCOL

1. Using a marking pen, divide the starch agar plate into three equal sectors. Be sure to mark on the bottom of the plate.

2. Spot inoculate two sectors with the test organisms leaving the third sector as a control.

3. Label the plate with the organisms' names, your name and the date.

4. Invert the plate and incubate it aerobically at 35°C for 48 hours.

5. Remove the plate from the incubator and flood it with Gram iodine. (See Precautions.)

6. Examine the plate for clearing around the bacterial growth.

7. Record your results in the space provided.

PRECAUTIONS

⚠ Gram Iodine is poisonous and should be treated with care.

⚠ Note the location and appearance of the growth *before* adding the iodine. Occasionally, growth that is thinning at the edge will give the appearance of clearing.

> **Photographic Atlas
> Reference
> Page 76**

REFERENCES

Collins, C. H., Patricia M. Lyne, J. M. Grange. 1995. Page 117 in *Collins and Lyne's Microbiological Methods, 7th Ed.* Butterworth-Heinemann, UK.

DIFCO Laboratories. 1984. Page 879 in *DIFCO Manual, 10th Ed.*, DIFCO Laboratories, Detroit, MI.

Lányi, B. 1987. Page 55 in *Methods in Microbiology, vol. 19*, edited by R. R. Colwell and R. Grigorova, Academic Press Inc., New York.

MacFaddin, Jean F. 1980. Page 286 in *Biochemical Tests for Identification of Medical Bacteria, 2nd Ed.* Williams & Wilkins, Baltimore, MD.

Smibert, Robert M. and Noel R. Krieg. 1994. Page 630 in *Methods for General and Molecular Bacteriology*, edited by Philipp Gerhardt, R. G. E. Murray, Willis A. Wood and Noel R. Krieg, American Society for Microbiology, Washington, DC.

TABLE OF RESULTS

RESULT	INTERPRETATION	SYMBOL
Clearing around growth	Amylase is present	+
No clearing around growth	Amylase is absent	−

ORGANISM	RESULT +/−

COMBINATION DIFFERENTIAL MEDIA

All the media included in this unit perform multiple tests and, as such, identify an organism's ability to perform (or not perform) specific *series'* of biochemical functions. Several examples of combination media are considered here. Triple sugar iron (TSI) agar and Kligler iron agar (KIA) are designed to differentiate enteric bacilli based on their ability to respire aerobically and anaerobically, and to ferment lactose. Litmus milk provides results for bacterial ability to ferment lactose, digest casein, coagulate casein, reduce litmus, and/or deaminate lactalbumin. Lysine iron agar (LIA) is used to differentiate enteric bacilli based on their ability to decarboxylate or deaminate lysine, and to respire anaerobically by reducing sulfur. SIM Medium (Sulfur-Indole-Motility) is used to detect the ability to reduce sulfur to H_2S, produce indole through the degradation of the amino acid tryptophan, and/or to demonstrate motility. Refer to the appropriate sections in the *Photographic Atlas* for more information.

Exercise 5–19	Kligler's Iron Agar (KIA)

MATERIALS

    Five KIA slants
    Recommended organisms (grown on solid media):
        Alcaligenes faecalis
        Citrobacter diversus
        Citrobacter freundii
        Proteus vulgaris

TEST PROTOCOL

Photographic Atlas Reference Page 56

1. Stab inoculate four KIA slants with the test organisms. Using a *heavy inoculum* stab the agar butt; then streak the slant.

2. Label the slants with the organisms' names, your name and the date.

TABLE OF RESULTS

RESULT	INTERPRETATION	SYMBOL
Red slant/red butt	No fermentation; peptone catabolized	K/K
Red slant/no change in butt	No fermentation; peptone used aerobically	K/NC
Red slant/yellow butt	Glucose fermentation; peptone catabolized	K/A
Yellow slant/yellow butt	Glucose and lactose fermentation	A/A
Cracks in or lifting of agar	Gas production	G+
Black in agar	Sulfur reduction (H_2S production)	H_2S+
No change in slant/no change in butt	Organism is either not growing at all or is growing slowly *and is not a member of the Enterobacteriaceae*	NC/NC

ORGANISM	RESULT

3. Incubate the slants aerobically with the uninoculated control at 35°C for 18 to 24 hours.

4. Examine the tubes for characteristic color changes and gas production.

5. Record your results in the space provided. Be sure to include results for the slant and butt *and* any gas or H$_2$S production. (See Figure 5-41 and Table 5-1 in the *Photographic Atlas*.)

PRECAUTIONS

⚠ Time is critical in this test; it must be read between 18 and 24 hours after inoculation.

⚠ The medium should be fresh. If it is more than 24 hours old, melt it in boiling water and reslant.

REFERENCES

Baron, Ellen Jo, Lance R. Peterson and Sydney M. Finegold. 1994. Chapters 10 in *Bailey and Scott's Diagnostic Microbiology, 9th Ed.* Mosby-Yearbook, St. Louis, MO.

DIFCO Laboratories. 1984. Page 485 in *DIFCO Manual, 10th Ed.* DIFCO Laboratories, Detroit, MI.

Koneman, Elmer W., Stephen D. Allen, William M. Janda, Paul C. Schreckenberger and Washington C. Winn, Jr. 1997. Chapter 4 in *Color Atlas and Textbook of Diagnostic Microbiology, 5th Ed.* J.B. Lippincott Company, Philadelphia, PA.

Lányi, B. 1987. Page 44 in *Methods in Microbiology, Vol. 19,* edited by R. R. Colwell and R. Grigorova, Academic Press Inc., New York.

MacFaddin, Jean F. 1980. Page 183 in *Biochemical Tests for Identification of Medical Bacteria, 2nd Ed.* Williams & Wilkins, Baltimore, MD.

Power, David A. and Peggy J. McCuen. 1988. Page 171 in *Manual of BBL® Products and Laboratory Procedures, 6th Ed.* Becton Dickinson Microbiology Systems, Cockeysville, MD.

Litmus Milk Medium

MATERIALS

Six litmus milk tubes
Recommended organisms:
  Bacillus megaterium
  Escherichia coli
  Lactobacillus acidophilus
  Morganella morganii
  Streptococcus faecium

TEST PROTOCOL

1. Inoculate five tubes with the test cultures and leave one tube uninoculated as a control.

2. Label the tubes with the organisms' names, your name and the date.

3. Incubate the tubes aerobically with the uninoculated control at 35°C for 7 to 14 days.

4. Examine the tubes for characteristic color changes, gas production, and clot formation.

5. Record your results in the space provided.

Photographic Atlas
Reference
Page 60

PRECAUTION

⚠ Because there are many possible reactions with this test, be sure to compare your results with an uninoculated control and with the table and photographs on pages 60 and 61 of the *Photographic Atlas*.

REFERENCES

MacFaddin, Jean F. 1980. Page 194 in *Biochemical Tests for Identification of Medical Bacteria, 2nd Ed.* Williams & Wilkins, Baltimore, MD.

Power, David A. and Peggy J. McCuen. 1988. Page 177 in *Manual of BBL® Products and Laboratory Procedures, 6th Ed.* Becton Dickinson Microbiology Systems, Cockeysville, MD.

TABLE OF RESULTS

RESULT*	INTERPRETATION	SYMBOL
Pink color	Acid production from lactose fermentation	A
White color (except top 1 cm)	Reduction of litmus	R
Pink and solid; clot not movable	Acid clot due to casein precipitation	AC
Semisolid and *not pink*; gray fluid at top	Rennet curd due to presence of rennin	C
Blue	Alkaline reaction due to degradation of milk protein	P
Fissures in clot	Gas production from lactose fermentation	G
Discoloration of medium; loss of "body"	Complete proteolysis of milk proteins	P
No change	None of the above reactions	NC

* NOTE: These results may be combined. See Figures 5-45, 5-46, and 5-47 in the *Photographic Atlas*.

ORGANISM	RESULT

Lysine Iron Agar (LIA)

MATERIALS

Six LIA slants

Recommended organisms (grown on solid media):
- *Citrobacter freundii*
- *Proteus mirabilis*
- *Shigella flexneri*
- *Escherichia coli*
- *Salmonella typhimurium*

TEST PROTOCOL

1. Stab inoculate five LIA slants with the test organisms. Using a heavy inoculum stab the agar butt twice; then, streak the slant.

2. Label the slants with the organisms' names, your name and the date.

3. Incubate the slants aerobically with the uninoculated control at 35°C for 18 to 24 hours.

4. Examine the tubes for characteristic color changes.

5. Record your results in the space provided.

Photographic Atlas Reference Page 62

REFERENCES

DIFCO Laboratories. 1984. Page 534 in *DIFCO Manual, 10th Ed.* DIFCO Laboratories, Detroit, MI.

Power, David A. and Peggy J. McCuen. 1988. Page 180 in *Manual of BBL® Products and Laboratory Procedures, 6th Ed.* Becton Dickinson Microbiology Systems, Cockeysville, MD.

TABLE OF RESULTS		
RESULT[1,2]	**INTERPRETATION**	**SYMBOL[3]**
Red slant	Lysine deaminase positive	red/
Purple slant	Lysine deaminase negative	purple/
Purple butt	Lysine decarboxylase positive	/purple
Yellow butt	Lysine decarboxylase negative	/yellow
Black butt	Sulfur reduction (H_2S production)	S
Cracks in or lifting of agar	Gas production	G

[1] Slants are read for deaminase activity and butts are read for decarboxylase activity.

[2] Yellow from decarboxylation may still be read if there is black coloration in the butt by holding the tube up to a light.

[3] Results are recorded with the color of the slant followed by the color of the butt (slant/butt). Gas and H_2S production follow. See Figure 5-51 in the *Photographic Atlas* for more information.

ORGANISM	RESULT

SIM Medium
(Sulfur Reduction Test, Indole Production, Motility)

MATERIALS

Four SIM tubes
Kovac's reagent (available commercially)
Recommended organisms:
    Escherichia coli
    Salmonella typhimurium
    Shigella flexneri

TEST PROTOCOL

Day One

1. Stab inoculate three SIM tubes with the test organisms leaving one tube uninoculated as a control.

2. Label the tubes with the organisms' names, your name and the date.

3. Incubate the tubes aerobically with the uninoculated control at 35°C for 24 to 48 hours.

Day Two

1. Examine the tubes for formation of black precipitate in the medium *and* spreading from the stab line. Using the information in the Tables of Results, record any H_2S production and/or motility in the space provided below.

2. Add a few drops of Kovac's reagent to each tube.

3. Observe for the formation of a red color in the reagent layer.

4. Record your results in the space provided.

PRECAUTION

 Kovac's reagent should be stored in the refrigerator. Although commercial brands include an expiration date, note the reagent's color. A color change from pale yellow to brown is an indication that the solution has deteriorated.

REFERENCES

DIFCO Laboratories. 1984. Page 762 in *DIFCO Manual, 10th Ed.* DIFCO Laboratories, Detroit, MI.

MacFaddin, Jean F. 1980. Page 162 in *Biochemical Tests for Identification of Medical Bacteria, 2nd Ed.* Williams & Wilkins, Baltimore, MD.

Power, David A. and Peggy J. McCuen. 1988. Page 246 in *Manual of BBL® Products and Laboratory Procedures, 6th Ed.* Becton Dickinson Microbiology Systems, Cockeysville, MD.

TABLE OF SULFUR REDUCTION RESULTS

RESULT	INTERPRETATION	SYMBOL
Black in the medium	Sulfur reduction (H_2S production)	+
No black in the medium	Sulfur is not reduced	−

TABLE OF MOTILITY RESULTS

RESULT	INTERPRETATION	SYMBOL
Growth spreading from the stab line	Motility	+
No spreading of growth	Nonmotile	−

TABLE OF INDOLE PRODUCTION RESULTS

RESULT	INTERPRETATION	SYMBOL
Red in the alcohol layer of Kovac's reagent	Tryptophan is converted to indole	+
Reagent color is unchanged	Tryptophan is not converted to indole	−

ORGANISM	H_2S +/−	INDOLE +/−	MOTILITY +/−

Exercise 5–23

Triple Sugar Iron (TSI) Agar

MATERIALS

Five TSI slants
Recommended organisms (grown on solid media):
- *Escherichia coli*
- *Proteus vulgaris*
- *Pseudomonas aeruginosa*
- *Salmonella typhimurium*

TEST PROTOCOL

1. Stab inoculate four TSI slants with the test organisms. Using a heavy inoculum stab the agar butt; then streak the slant. (See Precautions.)

2. Label the tubes with the organisms' names, your name and the date.

3. Incubate the slants aerobically with the uninoculated control at 35°C for 18 to 24 hours.

4. Examine the tubes for characteristic color changes and gas production.

5. Record your results in the space provided. Be sure to include results for the slant and butt *and* any gas or H_2S production. (See Figure 5-82 in the *Photographic Atlas*.)

PRECAUTION

⚠ Time is critical in this test; it must be read between 18 and 24 hours after inoculation.

Photographic Atlas Reference Page 80

REFERENCES

Baron, Ellen Jo, Lance R. Peterson and Sydney M. Finegold. 1994. Chapters 10 in *Bailey and Scott's Diagnostic Microbiology, 9th Ed.* Mosby-Yearbook, St. Louis, MO.

DIFCO Laboratories. 1984. Page 1019 in *DIFCO Manual, 10th Ed.* DIFCO Laboratories, Detroit, MI.

Koneman, Elmer W., Stephen D. Allen, William M. Janda, Paul C. Schreckenberger and Washington C. Winn, Jr. 1997. Chapter 4 in *Color Atlas and Textbook of Diagnostic Microbiology, 5th Ed.* J.B. Lippincott Company, Philadelphia, PA.

Lányi, B. 1987. Page 44 in *Methods in Microbiology, Vol. 19,* edited by R. R. Colwell and R. Grigorova, Academic Press Inc., New York.

MacFaddin, Jean F. 1980. Page 183 in *Biochemical Tests for Identification of Medical Bacteria, 2nd Ed.* Williams & Wilkins, Baltimore, MD.

Power, David A. and Peggy J. McCuen. 1988. Page 269 in *Manual of BBL® Products and Laboratory Procedures, 6th Ed.* Becton Dickinson Microbiology Systems, Cockeysville, MD.

TABLE OF RESULTS

RESULT	INTERPRETATION	SYMBOL
Red slant/red butt	No fermentation; peptone catabolized	K/K
Red slant/no change in butt	No fermentation; peptone used aerobically	K/NC
Red slant/yellow butt	Glucose fermentation; peptone catabolized	K/A
Yellow slant/yellow butt	Glucose and lactose and/or sucrose fermentation	A/A
Cracks in or lifting of agar	Gas production	G+
Black in agar	Sulfur reduction (H_2S production)	H_2S+
No change in slant/no change in butt	Organism is either not growing at all or is growing slowly *and is not a member of the Enterobacteriaceae*	NC/NC

ORGANISM	RESULT

ANTIBIOTIC SUSCEPTIBILITY TESTING

The therapeutic role of antibiotics in medicine is well established. Infections of all sorts have been treated successfully for decades by a wide variety of antimicrobial agents. The fact that these agents behave the way they do with certain organisms also makes them invaluable diagnostic tools. Bacteria that respond favorably to antibiotic treatment frequently show susceptibility to the same antimicrobial agents when tested *in vitro*. Research has demonstrated that many bacteria cannot only be identified using antibiotics, but to a certain degree, their virulence can be measured by their ability to resist them.

Although antibiotic susceptibility tests are performed in a variety of ways, all methods determine microbial resistance characteristics to a specific antibiotic. All employ the same demonstrable principle — that an antibiotic impregnated disk placed on an inoculated agar plate will produce a clearing in the growth of susceptible organisms and no clearing in the growth of resistant organisms. The bacitracin test included in this unit is typical of *diagnostic* susceptibility tests used to identify specific pathogens. The Kirby-Bauer test (included in Section Eight) is a test typically performed on known or suspected organisms to determine susceptibility patterns prior to prescribing a *therapeutic* course.

Exercise 5-24 — Bacitracin Susceptibility Test

MATERIALS

One blood agar plate (commercial preparation of TSA containing 5% sheep blood)
Sterile cotton applicators
0.04 unit bacitracin disks
Beaker of alcohol with forceps
Recommended organisms (grown in broth):
  Staphylococcus epidermidis
  Micrococcus roseus

TEST PROTOCOL

1. Inoculate half of the blood agar plate with *S. epidermidis*. Do this by making a single streak nearly halfway across the diameter of the plate. Turn the plate 90° and spread the organism evenly so as to produce a bacterial lawn covering half of the agar surface.

2. Being careful not to mix the organisms, repeat the process on the other half of the plate using *M. roseus*.

3. Sterilize the forceps by placing them in the Bunsen burner flame long enough to ignite the alcohol. Once the alcohol has burned off, use the forceps to place a bacitracin disk in the center of the *S. epidermidis* half of the plate. Gently tap the disk into place to prevent it from falling off when the plate is inverted.

Photographic Atlas Reference Page 38

4. Resterilize the forceps and place a bacitracin disk on the *M. roseus* half of the plate. Tap the disk into place and return the forceps to the alcohol.

5. Invert the plate, label it appropriately and incubate it for 24 to 48 hours at room temperature.

6. Remove the plate from the incubator and examine it for clearing around the disks.

7. Record your results in the space provided below. Refer to Section 5 of the *Photographic Atlas* for other applications of this test.

TABLE OF RESULTS

RESULT	INTERPRETATION	SYMBOL
Clearing zone around disk	Organism is sensitive to bacitracin	S
No clearing around disk	Organism is resistant to bacitracin	R

ORGANISM	RESULT +/−

PRECAUTIONS

⚠ Leave a small uninoculated space between the two organisms to avoid contaminating the cultures.

⚠ If necessary, allow the plate to dry slightly before inverting it to avoid mixing of the cultures.

⚠ Be careful to shield the top of the plate with the lid while you are inoculating it.

REFERENCES

Baron, Ellen Jo, Lance R. Peterson and Sydney M. Finegold. 1994. Page 329 in *Bailey & Scott's Diagnostic Microbiology, 9th Ed.* Mosby–Year Book, Inc. St. Louis, Missouri.

DIFCO Laboratories. 1984. Page 292 in *DIFCO Manual, 10th Ed.* DIFCO Laboratories, Detroit, MI.

Koneman, Elmer W., *et al.* 1997. Page 551 and 1299 in *Color Atlas and Textbook of Diagnostic Microbiology, 5th Ed.* Lippincott-Raven Publishers, Philadelphia, PA.

OTHER DIFFERENTIAL TESTS

This catch-all unit includes tests which do not fit elsewhere but are important tests to consider. The coagulase tests are commonly used to presumptively identify pathogenic *Staphylococcus* species. Motility agar is used to detect bacterial motility.

Exercise 5–25

Coagulase Tests (Slide)

MATERIALS

Commercially prepared dehydrated rabbit plasma
Sterile distilled or deionized water
Sterile saline
Sterile 1 mL pipettes
Microscope slides
Recommended organisms:
  Staphylococcus aureus
  Staphylococcus epidermidis

TEST PROTOCOL

1. Prepare two smears of each organism in sterile saline on microscope slides.

2. To one smear mix in a small drop of plasma; to the other smear add deionized water as a control.

3. Observe for clumping within 2 minutes.

4. Record your results in the space provided.

Photographic Atlas Reference Page 46

PRECAUTIONS

⚠ Use only fresh cultures to avoid false negative results.

⚠ Record slide test results after no more than 2 minutes to avoid false positive results.

TABLE OF RESULTS		
RESULT	**INTERPRETATION**	**SYMBOL**
Drop on slide shows visible clumping	Plasma has been coagulated	+
Drop is homogenous with no clumping	Plasma has not been coagulated	−

ORGANISM	RESULT +/−

Coagulase Tests (Tube)

MATERIALS

Three sterile test tubes (12 mm x 75 mm) containing sterile rabbit plasma
Sterile 1 mL pipettes
Microscope slides
Recommended organisms:
  Staphylococcus aureus
  Staphylococcus epidermidis

TEST PROTOCOL

1. Inoculate two plasma tubes with the test organisms and leave the third tube uninoculated as a control.

2. Label the tubes with the organisms' names, your name and the date.

3. Incubate the tubes with the uninoculated control at 35°C for up to 24 hours, checking for coagulation every 30 minutes for the first 2 to 4 hours.

4. Record your results in the table below.

PRECAUTIONS

⚠ Use only fresh cultures to avoid false negative results.

⚠ Do not run the test longer than 24 hours to avoid false negative results.

Photographic Atlas Reference Page 46

REFERENCES

Collins, C. H., Patricia M. Lyne, J. M. Grange. 1995. Page 111 in *Collins and Lyne's Microbiological Methods, 7th Ed.* Butterworth-Heinemann, UK.
DIFCO Laboratories. 1984. Page 232 in *DIFCO Manual, 10th Ed.* DIFCO Laboratories, Detroit, MI.
Holt, John G. (Editor). 1994. *Bergey's Manual of Determinative Bacteriology, 9th Ed.* Williams and Wilkins, Baltimore, MD.
Lányi, B. 1987. Page 62 in *Methods in Microbiology, Vol. 19,* edited by R. R. Colwell and R. Grigorova, Academic Press Inc., New York.
MacFaddin, Jean F. 1980. Page 64 in *Biochemical Tests for Identification of Medical Bacteria, 2nd Ed.* Williams & Wilkins, Baltimore, MD.

TABLE OF RESULTS		
RESULT	**INTERPRETATION**	**SYMBOL**
Medium is solid (within 24 hours)	Plasma has been coagulated	+
Medium is liquid (within 24 hours)	Plasma has not been coagulated	−

ORGANISM	RESULT +/−

| Exercise 5–27 | **Motility Test** |

MATERIALS

Three motility agar stabs
Recommended organisms:
  Enterobacter aerogenes
  Staphylococcus aureus

TEST PROTOCOL

1. Stab inoculate two tubes with the test organisms. Stab the third tube with a sterile inoculating wire as an uninoculated control.

2. Label the tubes with the organisms' names, your name and the date.

3. Incubate the tubes aerobically with the uninoculated control at 35°C for 24 to 48 hours.

4. Examine the growth pattern for characteristic spreading from the stab line.

5. Record your results in the space below.

PRECAUTION

⚠ The bacterial growth pattern can be obscured by careless stabbing technique (Figure 5-5). Therefore, *gently* pull the inoculating needle out of the agar exactly the way it went in.

REFERENCES

DIFCO Laboratories. 1984. Page 581 in *DIFCO Manual, 10th Ed.* DIFCO Laboratories, Detroit, MI.

MacFaddin, Jean F. 1980. Page 214 in *Biochemical Tests for Identification of Medical Bacteria, 2nd Ed.* Williams & Wilkins, Baltimore, MD.

Power, David A. and Peggy J. McCuen. 1988. Page 201 in *Manual of BBL® Products and Laboratory Procedures, 6th Ed.* Becton Dickinson Microbiology Systems, Cockeysville, MD.

> *Photographic Atlas Reference Page 64*

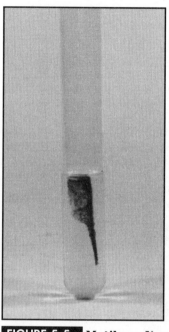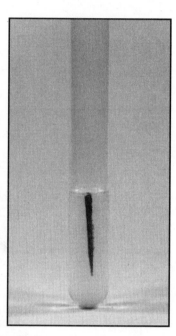

FIGURE 5-5 **Motile or Nonmotile?**

These photographs, taken of the same motility agar stab, demonstrate one possible result of poor inoculation technique. Because of its spreading appearance, the growth pattern shown on the left might be interpreted as motile. The growth pattern on the right, seen after rotating the tube 90°, demonstrates that the organism is nonmotile. In this example, the microbiologist moved the inoculating needle to the side as she removed it from the agar. True motility is demonstrated by growth radiating in all directions from the stab line.

TABLE OF RESULTS

RESULT	INTERPRETATION	SYMBOL
Reddish-purple extending diffusely from the stab line	The organism is motile	+
Reddish-purple only along the stab line	The organism is nonmotile	–

ORGANISM	RESULT +/–

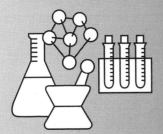

Determination of Bacterial Unknowns: Two Projects

*I*n this section you will be given a chance to demonstrate what you have learned in Sections One through Five. In fact you will be given two chances. The first exercise is a morphological unknown project. In it, you will use the skills you developed from the sections covering staining and microscopy. It will also acquaint you with the basic process of elimination typically used in microbial identification. The second unknown exercise is an extensive project which will take you several weeks to complete. It will give you the opportunity to apply all of the skills you have learned in this class to date. Enjoy these exercises; they can be great fun and very rewarding.

Exercise 6-1

Morphological Unknown

INTRODUCTION

In this exercise you will be given two pure bacterial cultures selected from the organisms listed in Table 6-1. Your job will be to identify each using only the staining techniques covered in Section Three. Perform *one stain at a time on each* and compare the results with those in the table. This process of eliminating the organisms whose characteristics do not match your results will eventually leave you with a single species — your unknown.

Your process of elimination will be only as good as your documentation of it. Therefore, you will record your progress in two ways: 1) by entering the necessary data in the spaces provided in Table 6-1 for each of your unknowns, and 2) by completing the flow charts in Figures 6-1 and 6-2. We have started the flow charts for you by giving you a few of the branches and listing appropriate organisms. As you complete each new stain and enter the results in the table you will need to add an additional branch to the appropriate flow chart. You will continue in this manner with *each* flow chart until you have eliminated all but one organism. The Procedure contains a detailed explanation of the process.

Take time to study the flow chart. It is simply an organized method of explaining the process of elimination which is the essence of unknown determination. As you can see, one group of organisms gets divided into two smaller groups based on each organism's reaction to a particular stain (shown in Table 6-1). For example, notice that the entire group of organisms can be divided based on individual Gram stain reactions; some are Gram-positive and the rest are Gram-negative. YOUR UNKNOWN IS ONE OF THE LISTED ORGANISMS. Therefore, when you perform a Gram stain on it, you will be able to identify the group it belongs in by *its* Gram reaction. Once you have determined which of the two smaller groups your organism belongs in, you can divide the smaller group again by performing another stain or noting its cell arrangement or morphology.

In short, each step of the process requires gathering data from the table in order to compare your organism's result to the list of possibilities. Each step of the process divides a group of remaining organisms into two smaller groups and adds another branch to the flow chart. This is done until all organisms are eliminated except for the unknown.

MATERIALS

Microscope
Clean microscope slides
Clean cover glasses
Gram stain kit
Acid fast stain kit
Capsule stain kit
Spore stain kit
Bunsen burner
Striker
Inoculating loop
Recommended organisms (other species of the same genera may be substituted): 18-24 hour nutrient broth cultures of
    Micrococcus roseus
    Enterococcus faecalis
    Staphylococcus aureus
    Bacillus cereus
    Lactobacillus plantarum
    Mycobacterium smegmatis
    Citrobacter diversus
    Vibrio harveyi
    Shigella flexneri
    Flavobacterium capsulatum

PROCEDURE

1. Obtain two unknown broth cultures. Record their numbers in the two boxes of Table 6-1 labeled "Unknown #" and above each flow chart in Figures 6-1 and 6-2. (This exercise may take two or more lab periods to complete. You should transfer your unknowns to nutrient agar slants at the end of each lab period to have fresh cultures for the next class. Incubate at 35°C.)

For convenience and clarity we have written the following instructions for one unknown. Simply apply the instructions to each.

---

2. Perform a Gram stain of the organism. The stain should be run with known Gram-positive and Gram-negative controls to verify your technique (see Figure 4-3 in the *Photographic Atlas*). The Gram stain will not only provide information on Gram reaction, but

will also allow observation of cell size, shape and arrangement. Enter all these in the appropriate columns of Table 6-1. (Note: cell size for the unknown candidates is not given and should not be used to differentiate the species. To give you an idea of typical cell sizes, the cocci should be about 1 µm in diameter and the rods 1 to 5 µm in length.)

3. Compare your result with the data in Table 6-1. Based on your result, follow the appropriate branch in the flow chart (Figures 6-1 or 6-2) until you reach the list of organisms that matches your unknown's Gram reaction and cell morphology.

4. Find a stain that will eliminate at least one of the remaining candidates. (Notice that this choice of subsequent stains will be dictated by the earlier results. For example, you wouldn't perform a spore stain on an organism already determined to be a coccus.) Continue the flow chart for each unknown by writing the remaining candidates' names under the appropriate flow chart branch. (Be sure to include all of the possible candidates. If you "lose" one due to a clerical error and it's your unknown, you've reduced your chance of successful identification to zero!)

5. Perform the stain and determine to which new branch of the flow chart your organism belongs. (Use the broths or the slants you inoculated as a source of organisms. Always use fresh cultures.)

6. Repeat Steps 4 and 5 until you eliminate all but one organism in the flow chart — this is your unknown!

7. After you identify your unknown, perform one more stain that you haven't done yet. This will act as a confirmatory test. If everything has gone correctly, you should get the predicted result (as given in Table 6-1) for your confirmatory test.

8. After the confirmatory test, write the name of the unknown in the appropriate space below. Have your instructor check your work. (If your confirmatory test result doesn't match the expected result, repeat any suspect stains and find the source of error.)

Unknown #_____ is _____

Unknown #_____ is _____

TABLE 6-1 **Table of Results for Organisms Used in this Exercise**
These results are typical for the species listed. Your particular strains may vary, due to their genetics, their age, or the environment in which they are grown.

ORGANISM	GRAM STAIN	CELL MORPHOLOGY	CELL ARRANGEMENT	CELL LENGTH (µm)	ACID-FAST STAIN	MOTILITY (WET MOUNT)	CAPSULE	SPORE STAIN
Micrococcus roseus	+	coccus	tetrads		−	−	−	
Enterococcus faecalis	+	coccus	pairs or chains		−	−	v	
Staphylococcus aureus	+	coccus	pairs or clusters		−	−		
Bacillus cereus	+	rod	single		−	+	−	
Lactobacillus plantarum	+	rod	short chains		−	−	−	
Mycobacterium smegmatis	weak +	rod	single or branched		+	−	−	
Citrobacter diversus	−	rod	single or pairs		−	+	−	
Vibrio harveyi	−	curved rod	single		−	+	−	
Shigella flexneri	−	rod	single		−	−	−	
Flavobacterium capsulatum	−	rod	single		−	−	+	
Unknown #								
Unknown #								

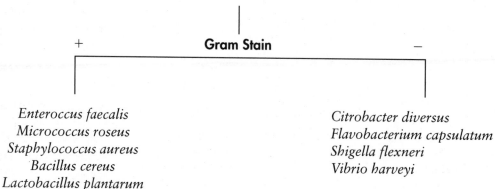

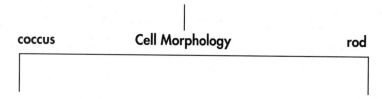

FIGURE 6-1 **Flow Chart for Unknown # _____**

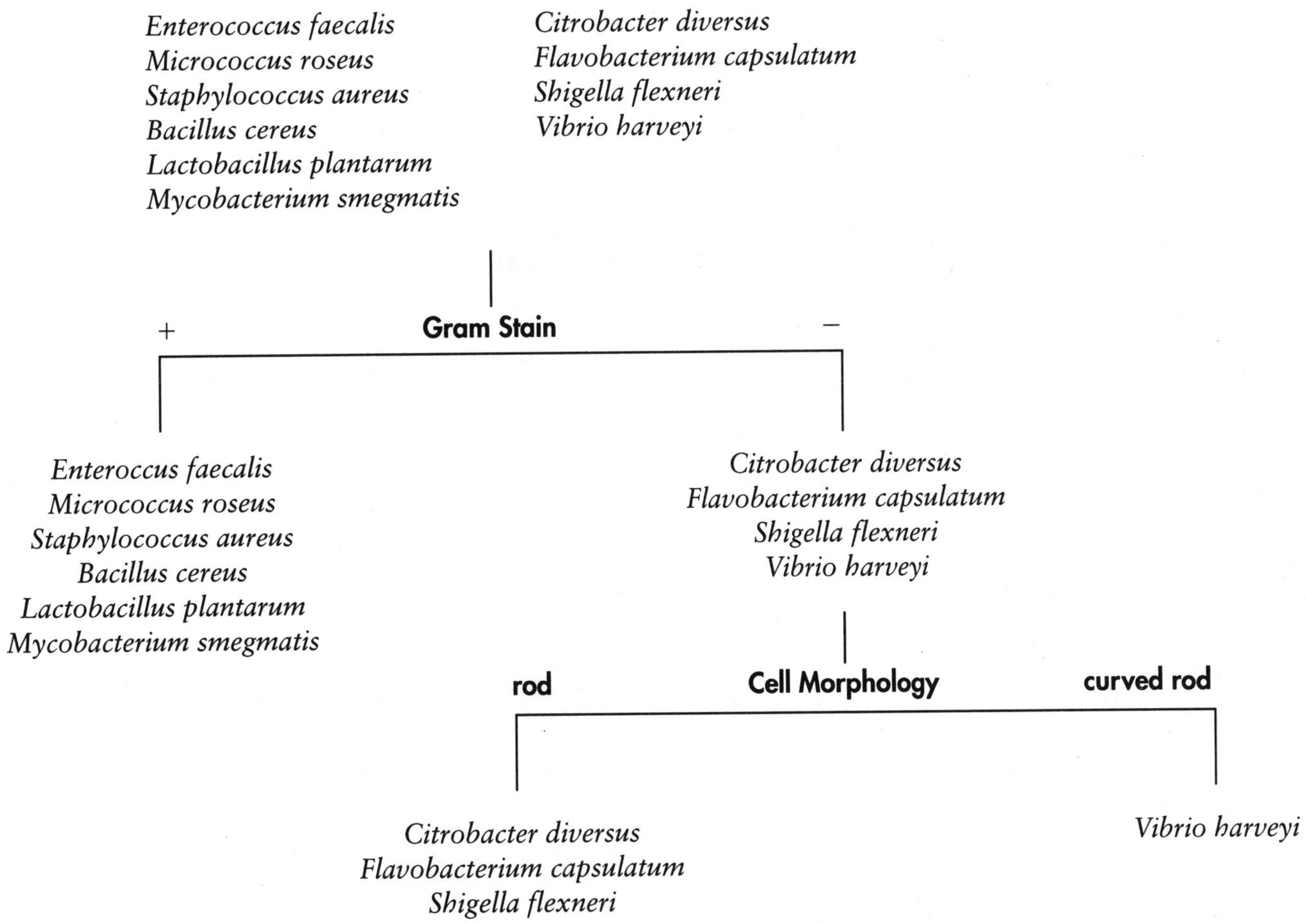

Enterococcus faecalis
Micrococcus roseus
Staphylococcus aureus
Bacillus cereus
Lactobacillus plantarum
Mycobacterium smegmatis

Citrobacter diversus
Flavobacterium capsulatum
Shigella flexneri
Vibrio harveyi

Gram Stain

+   −

Enteroccus faecalis
Micrococcus roseus
Staphylococcus aureus
Bacillus cereus
Lactobacillus plantarum
Mycobacterium smegmatis

Citrobacter diversus
Flavobacterium capsulatum
Shigella flexneri
Vibrio harveyi

rod   **Cell Morphology**   curved rod

Citrobacter diversus
Flavobacterium capsulatum
Shigella flexneri

Vibrio harveyi

FIGURE 6-2   **Flow Chart for Unknown # _____**

Bacterial Unknowns Project

INTRODUCTION

This is actually a term project to be done when you have completed sections One through Five. These sections prepare you with the skills needed to successfully do what professional laboratory microbiologists do daily: that is, *isolate* from mixed culture, *grow* in pure culture, and *identify* unknown species of bacteria. Scared? Don't worry. You will not be asked to perform any task not introduced in previous sections, so if you have any uncertainty regarding performance of these activities, refer to the appropriate exercises.

You will be given a mixture of two bacteria in broth culture. Your first job is to get the bacteria isolated and growing in pure culture using the techniques of Sections Two and Four. Next, Section Three techniques will be employed as you perform Gram stains on each to determine Gram reaction, and cell morphology and size. Lastly, results from the differential biochemical tests of Section Five will lead you to identification of your unknowns. And, of course, all work must be done safely and aseptically using the methods covered in Lab Safety and Section One.

As in Exercise 6-1 (the Morphological Unknown), you will be expected to keep accurate records of all activities, including mistakes and unexpected or equivocal results. (Remember that honesty is better than flawless technique in science. In the "real world," patients or an employer may be relying on your data being current and an accurate reflection of what actually was done.) You also will be expected to construct a flow chart on the back of each record sheet to show the tests you ran and the organisms eliminated by each until you have eliminated all but your unknown. The flow chart is a visual presentation of your thought processes in solving this problem.

While this project may seem intimidating, it is manageable if you don't try to do too much too fast. Take your time and think about what you want to do next based on earlier results. Employ lessons learned in Exercise 6-1 and closely follow the procedures listed below. Our experience tells us that students traditionally do one of three things with an unknown project:

- They treat it as a fun puzzle to solve
- They become completely stressed out and hate every minute of it, or
- They get lost early on and, for whatever reason, never quite catch up.

Please, for your own sake, ask for help if you need it. This exercise gives you a unique opportunity to use your mind in a slightly different way than other educational experiences. Take advantage of it; there is great fun and satisfaction to be had if you do. Happy hunting!

MATERIALS

Recommended organisms (to be mixed in pairs in a broth by the lab technician immediately prior to use):

Gram-positives
Bacillus cereus
Bacillus coagulans
Corynebacterium xerosis
Enterococcus faecalis
Lactobacillus plantarum
Micrococcus luteus
Micrococcus roseus
Mycobacterium smegmatis
Staphylococcus aureus
Staphylococcus epidermidis
Streptococcus lactis

Gram-negatives
Aeromonas hydrophila
Alcaligenes faecalis
Chromobacterium violaceum
Citrobacter amalonaticus
Enterobacter aerogenes
Erwinia amylovora
Escherichia coli
Hafnia alvei
Morganella morganii
Proteus mirabilis
Pseudomonas aeruginosa

PROCEDURE

1. Preliminary duties

   a. Your instructor will tell you which organisms will actually be used based on your lab's inventory. You will also be advised as to the optimum temperature for each organism.

   b. Your instructor will tell you which media and stains will be available for testing.

   c. Working as a class, you will determine the results of each organism for each test available. This information will provide a data base of results that you can use to compare against your unknown's results. We recommend running and using these "class controls" in lieu of referring to results in *Bergey's*

Manual of Systematic Bacteriology or some other standard reference for the following reasons

- as much as 10% of the strains of a species listed as positive or negative on a test give the opposite result
- many species have even higher strain variability
- not all test results are listed for all organisms

d. Class control tests should be run for the standard times, unless the timing of class sessions makes this impractical. It is imperative that incubation times for tests on unknowns be exactly the same as for the controls.

e. Incubate your class controls at the optimum temperature for each species as given to you by your instructor.

f. Class control results will be collected and distributed to each student. Your instructor will provide details on this process.

2. Isolation of the Unknown

a. You will be given a broth containing a *fresh* mixture of two unknown bacteria selected from the list of possible organisms. Enter the number of your unknown on your record sheets.

b. Mix the broth, then streak for isolation on two agar plates. Your instructor will tell you what media are available for use. You may be supplied with an undefined medium, such as trypticase soy agar, or a selective medium, such as phenylethyl alcohol agar or desoxycholate agar. Enter all relevant information concerning your isolation procedures on the record sheets, including date, medium, source of bacteria, and incubation temperature.

c. Incubate one agar plate at 25°C and the other at 37°C for at least 24 hours.

d. After incubation, check for isolated colonies that have different morphologies. If you have isolation of both, go on to step "2f." If you do *not* have isolation of *either*, continue with step "2e." If you have isolation of only one, go to step "2f" for the isolated species and step "2e" for the one not isolated. Be sure to enter relevant information on the record sheets.

e. If you do not have isolation, follow the advice that best matches your situation.

- Look for evidence of different growth, even though they're not separate. If you see different growth, use a portion of each and streak more plates. Then incubate them for either a shorter

time or at a suboptimal temperature, since they grew *too* well the first time.

- If you have only one type of growth, ask for a selective medium that favors growth of the one you're missing. Streak the mixture and incubate again. Also, reincubate your original plate — some species are slow growers, so their absence may be due to a slow growth rate.

After incubation, observe the plate(s) and repeat or go to step "2f," whichever is appropriate. You may also consult with your instructor for guidance in particularly difficult situations.

f. Once you have isolation, transfer a portion of each different colony type onto an agar slant to produce a pure culture. Use the rest of each colony for a Gram stain. After staining, label your pure cultures accordingly. Enter the following information on your record sheet: optimum growth temperature, colony morphology, cell morphology and arrangement, and cell size.

3. Identification of the Unknown

a. Follow this procedure for each unknown. Be sure to enter inoculation and reading dates for each test on the record sheet. Also include the test result and any comments about the test that explain any deviation from standard procedure (*i.e.,* reading tests before or after the accepted incubation time, running a test and not using it in the flow chart, *etc.*).

b. Construct a flow chart that divides all the organisms up first by Gram reaction, then by cellular morphology. Use the flow chart in Figure 6-1 as a style guide even though the actual organisms will be different.

c. Find the group of organisms that matches the results of your unknown. Then, look at your class controls results for *just those organisms* and choose a test that will divide these organisms into at least two groups. (Also consider your ability to return and read the test after the appropriate incubation time. That is, don't inoculate a 48 hour test on Thursday if you can't get into the lab on Saturday!) Continue the flow chart from the branch with the remaining candidates for your unknown.

d. Inoculate the test medium you have chosen and incubate it for the appropriate time. Your incubation temperature will be determined by the optimum temperature for growth on your streak plates. *It is important to run your tests at that optimum temperature because this is the temperature at which the class controls were run.*

e. While the test is incubating, you should begin planning what your next test will be. A final decision about the next test cannot be made until you have results from the first, but you can decide which test to run if the result is positive, and which one to run if it is negative. This applies to all stages of the flow chart: since you won't know if a subsequent test is relevant or not until you have results from the current one, *you should not inoculate a medium until you have those results*. It's really easy: run one test at a time for each unknown. (Note: In a clinical situation, where rapid identification of a pathogen may save a patient's life, tests are routinely run concurrently. However, remember that correct identification is only one objective of this project. More important is for you to demonstrate an understanding of the logic behind the process and execute it in the most efficient manner.)

f. Repeat the process of inoculating a medium, getting the results, and then choosing a subsequent test until you eliminate all but one organism. This *should* be your unknown.

g. When you have eliminated all but one organism, you will run one more test — the confirmatory test. This must be one that has not been run previously on your organism. It's also nice (but not necessary) if the test you choose gives a positive result (since, in general, we have more confidence in positive results than in negative results, because false positives are harder to get than false negatives.) The confimatory test provides you with the unique opportunity to predict the result before you run the test. If it matches, you are more certain that you have correctly identified your unknown. Continue with step "3j." If it doesn't match, continue with step "3h."

h. If your confirmatory test doesn't match the result you expected for your unknown, check with your instructor to see where your organism was eliminated incorrectly. In most cases, it will be difficult to determine at this point what was responsible — you, the class controls, or the organism itself.

Misidentification may be due to one or any combination of factors:

- the test procedure could have been done incorrectly by you or the person responsible for running class controls on your organism
- the test may have been interpreted incorrectly by you or the person responsible for running class controls on your organism
- the inoculum was too small in your test or the class controls to give a postive result in the limited incubation time
- the wrong organism was inoculated at the time of the test or the class controls (they all look alike in a tube, and once the label goes on, as far as the microbiologist is concerned, that culture becomes the labeled organism regardless of what's really in there!)
- for whatever reason, the organism just didn't react "correctly" during your test or during the class controls.

i. Based on your instructor's advice, you will either
- rerun the test where you incorrrectly eliminated your unknown (and perhaps rerun the test on the remaining organisms to check the class control results) or
- rerun the test where you incorrectly eliminated your unknown without running the controls again, or
- eliminate the problematic test from your flow chart, but use the other tests you've already done. If these don't allow you to identify your unknown, more tests will need to be run. Continue with step "3f."

j. When you have correctly identified your unknown, complete the record sheet and turn it in. Your instructor will advise you as to the point value of each section, the grading scale, and any other items that are required.

Unknown Number _____   **Student Name** _____

Isolation Procedure _____

_____

_____

_____

_____

_____

_____

Colony Morphology _____

Gram Stain _____   **Morphology and Arrangement** _____

Cell Dimensions _____   **Optimum Temperature** _____

Differential Tests

Test #1: _____   Date Begun: _____   Date Read: _____   Result: _____
Comments: _____

Test #2: _____   Date Begun: _____   Date Read: _____   Result: _____
Comments: _____

Test #3: _____   Date Begun: _____   Date Read: _____   Result: _____
Comments: _____

Test #4: _____   Date Begun: _____   Date Read: _____   Result: _____
Comments: _____

Test #5: _____   Date Begun: _____   Date Read: _____   Result: _____
Comments: _____

Test #6: _____   Date Begun: _____   Date Read: _____   Result: _____
Comments: _____

Test #7: _____   Date Begun: _____   Date Read: _____   Result: _____
Comments: _____

Test #8: _____   Date Begun: _____   Date Read: _____   Result: _____
Comments: _____

Test #9: _____   Date Begun: _____   Date Read: _____   Result: _____
Comments: _____

Test #10: _____   Date Begun: _____   Date Read: _____   Result: _____
Comments: _____

Test #11: _____   Date Begun: _____   Date Read: _____   Result: _____
Comments: _____

Test #12: _____   Date Begun: _____   Date Read: _____   Result: _____
Comments: _____

My Unknown is: _____

Unknown Number _____   **Student Name** _____

Isolation Procedure _____

_____

_____

_____

_____

_____

_____

_____

Colony Morphology _____

Gram Stain _____   **Morphology and Arrangement** _____

Cell Dimensions _____   **Optimum Temperature** _____

Differential Tests

Test #1: _____   Date Begun: _____   Date Read: _____   Result: _____
Comments: _____

Test #2: _____   Date Begun: _____   Date Read: _____   Result: _____
Comments: _____

Test #3: _____   Date Begun: _____   Date Read: _____   Result: _____
Comments: _____

Test #4: _____   Date Begun: _____   Date Read: _____   Result: _____
Comments: _____

Test #5: _____   Date Begun: _____   Date Read: _____   Result: _____
Comments: _____

Test #6: _____   Date Begun: _____   Date Read: _____   Result: _____
Comments: _____

Test #7: _____   Date Begun: _____   Date Read: _____   Result: _____
Comments: _____

Test #8: _____   Date Begun: _____   Date Read: _____   Result: _____
Comments: _____

Test #9: _____   Date Begun: _____   Date Read: _____   Result: _____
Comments: _____

Test #10: _____   Date Begun: _____   Date Read: _____   Result: _____
Comments: _____

Test #11: _____   Date Begun: _____   Date Read: _____   Result: _____
Comments: _____

Test #12: _____   Date Begun: _____   Date Read: _____   Result: _____
Comments: _____

My Unknown is: _____

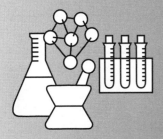

Quantitative Techniques

Microbiological quantitative techniques, simply defined, are the methods used to estimate the number of microorganisms or viruses per unit volume (*density*) in a given specimen. Of the five techniques examined in this section, only one ("Direct Count") involves the actual counting of cells. The other four employ a variety of simple techniques to *indirectly* measure density. The turbidometric technique used in the last exercise, Bacterial Growth in a Closed System, may be used for quantitative purposes, but we employ it as a simple and effective way of examining the growth of a bacterial population in a system where no nutrients are added beyond what is in the original medium and no wastes are removed. The information will be used to construct a growth curve.

As you proceed through the exercises in this section you will see the recurring item *serial dilution*. In microbiology, as well as other areas of science, a working knowledge of serial dilutions and dilution factors is essential. Although the math may seem difficult at first it will become easier with time and practice. Ask your instructor to give you practice problems to help you become proficient with the formulas.

In preparation for the exercises contained in this section, study the theoretical material and photographs in Section Six of the *Photographic Atlas* or your text. Also read through each exercise and discuss it with your lab partners in time to ask your instructor for help if there are aspects you don't understand. Each of the techniques you will be performing contains inherent strengths and weaknesses. Try to be as exact as you can in your measurements to maximize these strengths and produce reliable results.

Viable (Plate) Count

MATERIALS

Sterile 1 mL pipettes and pipettor
Two 9.9 mL dilution tubes (Appendix B)
Three 9.0 mL dilution tubes (Appendix B)
Eight nutrient agar plates
Beaker containing ethanol and a bent glass rod
Hand tally counter
Colony counter
24 hour broth culture of *Eschericia coli*

TEST PROTOCOL (See Figure 7-1)

The following procedure includes inoculation by the "spread plate technique." For instruction on this technique refer to Section One, Figures 1-8 and 1-11.

Day One

1. Organize the plates into 4 pairs and label them A, B, C and D.

2. Aseptically transfer 0.1 mL from the broth culture (the *original sample*) to a 9.9 mL dilution tube and mix well. This is *DF* (dilution factor) 10^{-2}. (For a discusion of dilution factors, see Precautions and pages 83–85 of the *Photographic Atlas*.)

3. Aseptically transfer 0.1 mL DF 10^{-2} to a 9.9 mL dilution tube. Mix well. This is *DF 10^{-4}*.

4. Aseptically transfer 1.0 mL DF 10^{-4} to a 9.0 mL dilution tube. Mix well. This is *DF 10^{-5}*.

5. Aseptically transfer 1.0 mL DF 10^{-5} to a 9.0 mL dilution tube. Mix well. This is *DF 10^{-6}*.

6. Aseptically transfer 1.0 mL DF 10^{-6} to a 9.0 mL dilution tube. Mix well. This is *DF 10^{-7}*.

7. Aseptically transfer 0.1 mL DF 10^{-4} to plate A. Using the spread plate technique disperse the sample evenly over the entire surface of the agar. Repeat with the second plate A and label both plates "0.1 mL of DF 10^{-4}" (the volume transferred to the plate *and* the dilution factor of the tube it came from).

8. Following the same procedure, transfer 0.1 mL volumes from DF 10^{-5}, DF 10^{-6} and DF 10^{-7} to plates B, C and D respectively. Label the plates accordingly.

9. Invert the plates and incubate at 35°C for 24 to 48 hours.

Day Two

Photographic Atlas Reference Page 83

1. After incubation set the countable plates aside (plates with 30 to 300 colonies) and properly dispose of the uncountable plates. Only one pair of plates should be countable.

2. Count the colonies on both plates and calculate the average. (Figure 7-2)

3. Using the formula below, determine the cell density of the original sample Note: "#CFU" (Colony Forming Units) replaces "colonies counted" in the formula.

$$\text{Original cell density} = \frac{\#CFU}{(\text{mL plated})\,(\text{dilution factor})}$$

Original cell density =

PRECAUTIONS

⚠ Dilution factors in serial dilutions are calculated using the following formula: $D_2 = (V_1)(D_1)/V_2$ where D_2 is the new dilution factor being calculated, D_1 is the dilution factor of the sample being diluted (undiluted samples have a dilution factor of 1), V_1 is the volume of sample to be diluted and V_2 is the combined volume of sample and diluent. For example, to calculate the new dilution factor when adding 0.1 mL of DF 10^{-2} to 9.9 mL of diluent (producing a combined volume of 10.0 mL), use the formula as illustrated below.

$$D_2 = \frac{(V_1)(D_1)}{V_2}$$

$$D_2 = \frac{(0.1\ \text{mL})(10^{-2})}{10.0\ \text{mL}} = 10^{-4}$$

⚠ By convention, when 0.1 mL of a sample is inoculated onto a plate, the dilution factor is recorded as tenfold greater and the "volume plated" term in the equation is omitted. This is because 0.1 mL has one-tenth the cells of 1.0 mL, which is comparable to another tenfold dilution at the time of plating. For example, if 0.1 mL of DF 10^{-2} is plated, then the DF on the plate (*the final dilution factor*) is recorded as FDF 10^{-3} and the volume plated is ignored since it has already been

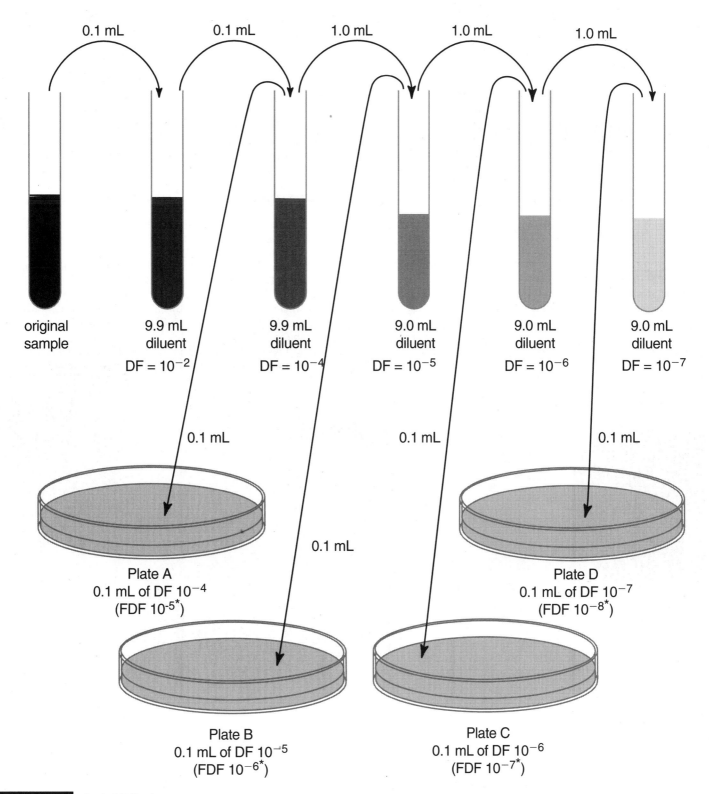

FIGURE 7-1   **Serial Dilution**

This is an illustration of the dilution scheme outlined in the test protocol. Use it as a guide while following the outlined procedure. As you perform the series of dilutions you will be assigning "dilution factors" to the tubes and their contents. These dilution factors do not indicate the concentration of cells in any of the tubes but how much the original sample has been diluted thus far. The number of colonies formed on the plates will provide the needed information to calculate the original cell concentration. Use the formulas provided in the Test Protocol section to do your calculations. For an explanation of dilution factors refer to the Precautions section.

*FDF *(Final dilution factor)* incorporates the 0.1 mL aliquot as if it were an additional tenfold dilution. It is offered as *an option only* to be used with the shortcut formula in Precautions.

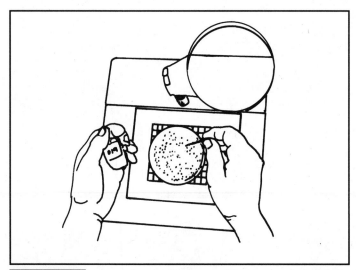

FIGURE 7-2 **Counting Bacterial Colonies**
Place the open plate on the colony counter, turn on the light and adjust the magnifying glass until all the colonies are visible. Using the grid in the background as a guide, count the colonies one section at a time. Mark each colony with a toothpick (to avoid counting it more than once) as you record with a hand tally counter. Properly dispose of the plate and toothpick when finished.

taken into account. The original density can then be calculated using the following formula.

$$\text{Original cell density} = \frac{\#CFU}{\text{Dilution factor}}$$

(It should be noted that this formula is a shortcut and will not provide the necessary units. If you choose to use this formula instead of the full version on page 112, do not forget to record your results in CFU/mL.)

REFERENCES

Collins, C.H., Patricia M. Lyne, J.M. Grange. 1995. Page 149 in *Collins and Lyne's Microbiological Methods, 7th Ed.* Butterworth-Heinemann, UK.

DIFCO Laboratories. 1984. Page 619 in *DIFCO Manual, 10th Ed.* DIFCO Laboratories, Detroit, MI.

Koch, Arthur L. 1994. Page 254 in *Methods for General and Molecular Bacteriology*, edited by Philipp Gerhardt, R.G.E. Murray, Willis A. Wood and Noel R. Krieg, American Society for Microbiology, Washington, DC.

Postgate, J.R. 1969. Page 611 in *Methods in Microbiology, Vol. 1.*, edited by J.R. Norris and D.W. Robbins, Academic Press, Inc., New York.

Direct Count

MATERIALS

Petroff-Hausser counting chamber with coverslip (Figure 7-3)
Test tubes
1 mL pipettes with pipettor
Pasteur pipettes with bulbs
Hand counter
Staining agents A and B (Appendix B)
Recommended organism (grown in broth):
Proteus vulgaris

TEST PROTOCOL

1. Transfer 0.1 mL from the original 24 hour culture tube to a non-sterile test tube.

2. Add 0.4 mL Agent A and 0.5 mL Agent B and mix well. (This dilution may not be suitable in all situations. Adjust the proportions of the broth culture and agents A and B if necessary to obtain a countable dilution, but remember to keep the total solution volume at 1.0 mL for easier calculation later. There should be at least five cells per small square to be a countable dilution.)

3. Place a coverslip on the Petroff-Hausser counting chamber and add a drop of the mixture such that it fills the well under the coverslip by capillary action.

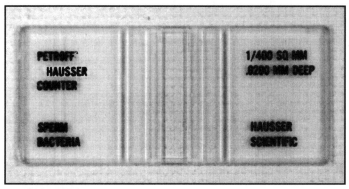

FIGURE 7-3   **Petroff-Hausser Counting Chamber**
The Petroff-Hausser counting chamber is a device used for the direct counting of bacterial cells. To examine a bacterial broth or suspension, place a drop of the sample in the chamber, cover it with a coverslip and place it on the microscope stage. Locate the grid in the center using the *low power* objective lens. Do not increase the magnification until you have found the grid on low power. Increase the magnification, focusing one objective at a time, until you have the cells and the grid in focus with the oil immersion lens. Follow the instructions in the test protocol.

4. Observe in the microscope and count the number of cells above each of five small squares.

Photographic Atlas Reference Page 86

5. Take the average and compute the original cell density using the formula below. Record your results in the space provided. (For an explanation of dilution factor calculations, see Precaution below.)

$$\text{Original cell density} = \frac{\text{Cells/small square}}{(\text{Volume})(\text{Dilution factor})}$$

$$\text{Original cell density} = \frac{\text{Cells/small square}}{(5 \times 10^{-8}\ \text{mL})(10^{-1})}$$

Original cell density =

PRECAUTION

⚠ Dilution factor is calculated with the following formula: $D_2 = (V_1)(D_1)/V_2$ where D_2 is the new dilution factor being calculated, D_1 is the dilution factor of the sample being diluted (undiluted samples have a dilution factor of 1), V_1 is the volume of sample to be diluted and V_2 is the combined volume of sample and diluent. The dilution factor for #2 in the test protocol above is calculated as follows:

$$D_2 = \frac{(V_1)(D_1)}{V_2}$$

$$D_2 = \frac{(0.1\ \text{mL})(1)}{1.0\ \text{mL}} = 10^{-1}$$

REFERENCES

Koch, Arthur L. 1994. Page 251 in *Methods for General and Molecular Bacteriology*, edited by Philipp Gerhardt, R. G. E. Murray, Willis A. Wood and Noel R. Krieg, American Society for Microbiology, Washington, DC.

Postgate, J. R. 1969. Page 611 in *Methods in Microbiology*, vol. 1., edited by J. R. Norris and D. W. Ribbons, Academic Press, Inc., New York.

Exercise 7–3

Plaque Assay For Determination of Phage Titer

MATERIALS

1 mL sterile pipettes
Six 9.0 mL and one 9.9 mL dilution tubes
Six nutrient agar plates
Six soft agar tubes
Hot water bath set at 45°C
T4 coliphage (Carolina Biological #K3-12-4330)
24 hour broth culture of *Escherichia coli* B (T-series phage host — Carolina Biological #K3-12-4300)

TEST PROTOCOL (Figure 7-4)

This procedure presumes the original phage titer to be between 1.0×10^6 and 1.0×10^{11} PFU (Plaque Forming Units) /mL. If the sample you receive is not in this range, you will need to adjust your dilution scheme up or down, accordingly, to produce countable plates. (For further explanation of dilution factors and serial dilutions see Precautions below and Figure 7-1 in Viable Count. See also pages 83–87 of the *Photographic Atlas*.)

1. Aseptically transfer 0.1 mL of the original sample to a 9.9 mL dilution tube. Mix well. This is *DF* (dilution factor) 10^{-2}.

2. Aseptically transfer 1.0 mL DF 10^{-2} to a 9.0 mL dilution tube. Mix well. This is *DF 10^{-3}*.

3. Aseptically transfer 1.0 mL DF 10^{-3} to a 9.0 mL dilution tube. Mix well. This is *DF 10^{-4}*.

4. Continue in this manner (adding 1.0 mL to 9.0 mL) until you have completed DF 10^{-8}.

5. Remove one soft agar tube from the hot water bath and add 0.1 mL DF 10^{-3} and 0.3 mL of the *E. coli* broth culture. Mix well and immediately pour onto a nutrient agar plate. Gently tilt the plate back and forth until the soft agar mixture is spread evenly across the solid medium. Label the plate "0.1 mL of DF 10^{-3}" (the volume transferred to the plate and the dilution factor of the tube it came from).

6. Repeat this procedure with dilutions 10^{-4}, 10^{-5}, 10^{-6}, 10^{-7}, and 10^{-8}. Label the plates accordingly.

7. Allow the agar to solidify completely.

8. Invert the plates and incubate, aerobically, at 35°C for 24 to 48 hours.

9. After incubation count the plaques (Figure 7-5).

10. Determine the original phage titer using the formula below. Record your results in the space provided. #PFU (plaque forming units) replaces "plaques counted" in the formula.

Photographic Atlas Reference Page 87

$$\text{Original phage density} = \frac{\text{\#PFU}}{(\text{mL plated})(\text{Dilution factor})}$$

Original phage density =

PRECAUTIONS

⚠ The soft agar tubes will solidify rather quickly upon removal from the hot water bath. Be prepared to use them immediately.

⚠ Prewarming the nutrient agar plates in the incubator will slow down the solidifying of the soft agar and allow more time for it to spread evenly over the solid medium.

⚠ Dilution factors in serial dilutions are calculated using the following formula: $D_2 = (V_1)(D_1)/V_2$ where D_2 is the new dilution factor being calculated, D_1 is the dilution factor of the sample being diluted (undiluted samples have a dilution factor of 1.0), V_1 is the volume of sample to be diluted and V_2 is the combined volume of sample and diluent. For example, to calculate the new dilution factor when adding 0.1 mL of DF 10^{-2} to 9.9 mL of diluent, use the formula thus:

$$D_2 = \frac{(V_1)(D_1)}{V_2}$$

$$D_2 = \frac{(0.1 \text{ mL})(10^{-2})}{10.0 \text{ mL}} = 10^{-4}$$

⚠ By convention, when 0.1 mL of a sample is inoculated onto a plate, the dilution factor is recorded as tenfold greater and the "volume plated" term in the equation is omitted. This is because 0.1 mL has one-tenth the phages of 1.0 mL, which is comparable to another tenfold dilution at the time of plating. For example, if 0.1 mL of DF 10^{-2} is plated, then the DF on the plate (*the final dilution factor*) is recorded as FDF 10^{-3} and the volume plated is ignored since it has already been

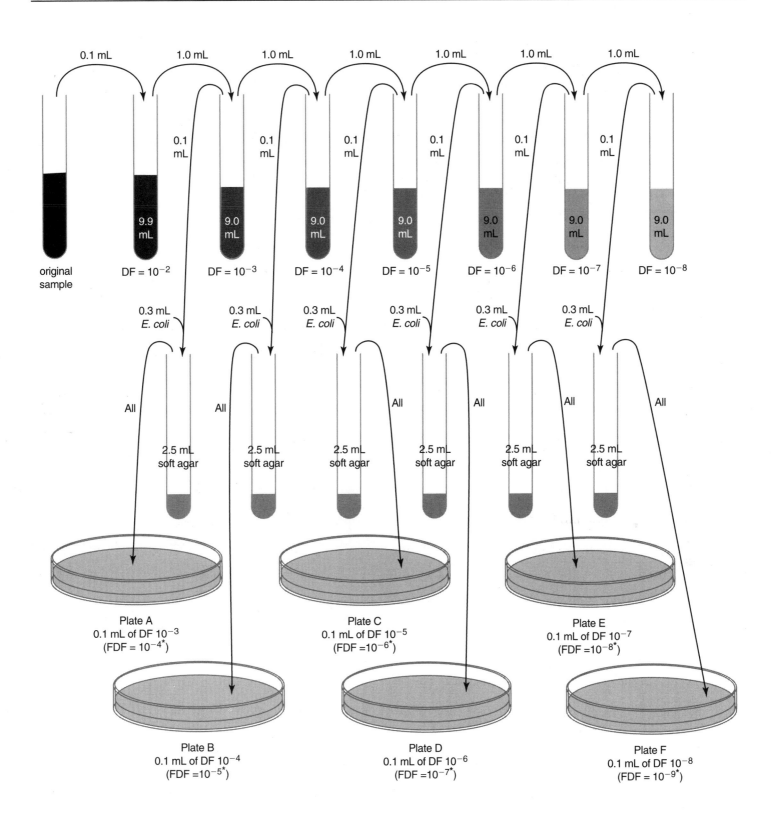

0.1 mL   1.0 mL   1.0 mL   1.0 mL   1.0 mL   1.0 mL   1.0 mL

0.1 mL   0.1 mL   0.1 mL   0.1 mL   0.1 mL   0.1 mL

original
sample

9.9 mL   9.0 mL   9.0 mL   9.0 mL   9.0 mL   9.0 mL   9.0 mL

$DF = 10^{-2}$ $DF = 10^{-3}$ $DF = 10^{-4}$ $DF = 10^{-5}$ $DF = 10^{-6}$ $DF = 10^{-7}$ $DF = 10^{-8}$

0.3 mL   0.3 mL   0.3 mL   0.3 mL   0.3 mL   0.3 mL
E. coli   *E. coli*   *E. coli*   *E. coli*   *E. coli*   *E. coli*

All   All   All   All   All   All

2.5 mL   2.5 mL   2.5 mL   2.5 mL   2.5 mL   2.5 mL
soft agar   soft agar   soft agar   soft agar   soft agar   soft agar

Plate A
0.1 mL of DF 10^{-3}
(FDF = 10^{-4*})

Plate C
0.1 mL of DF 10^{-5}
(FDF = 10^{-6*})

Plate E
0.1 mL of DF 10^{-7}
(FDF = 10^{-8*})

Plate B
0.1 mL of DF 10^{-4}
(FDF = 10^{-5*})

Plate D
0.1 mL of DF 10^{-6}
(FDF = 10^{-7*})

Plate F
0.1 mL of DF 10^{-8}
(FDF = 10^{-9*})

FIGURE 7-4   **Procedural Diagram for Plaque Assay**
Use this diagram as a guide while performing the dilutions outlined in Test Protocol.

*FDF *(Final dilution factor)* incorporates the 0.1 mL aliquot as if it were an additional tenfold dilution. It is offered as *an option only* to be used with the shortcut formula in Precautions.

FIGURE 7-5 **Counting Plaques**

Place the inverted plate on the colony counter, turn on the light and adjust the magnifying glass until all the plaques are visible. Using the grid in the background as a guide, count the plaques one section at a time. Mark each plaque with a felt tip marker (to avoid counting it more than once) as you record it with a hand tally counter.

taken into account. The original density can then be calculated using the following formula.

$$\text{Original phage density} = \frac{\#PFU}{\text{Dilution factor}}$$

(It should be noted that this formula is a shortcut and will not provide the necessary units. If you choose to use this formula instead of the full version on page 116, do not forget to record your results in PFU/mL.)

REFERENCES

Collins, C. H., Patricia M. Lyne, J. M. Grange. 1995. Page 149 in *Collins and Lyne's Microbiological Methods, 7th Ed.* Butterworth-Heinemann, UK.

DIFCO Laboratories. 1984. Page 619 in *DIFCO Manual, 10th Ed.* DIFCO Laboratories, Detroit, MI.

Province, David L. and Roy Curtiss III. 1994. Page 328 in *Methods for General and Molecular Bacteriology*, edited by Philipp Gerhardt, R. G. E. Murray, Willis A. Wood and Noel R. Krieg, American Society for Microbiology, Washington, DC.

Exercise 7–4 — Semiquantitative Streak of a Urine Specimen

MATERIALS

One blood agar plate (TSA with 5% sheep blood)
One sterile volumetric inoculating loop (either 0.01 mL or 0.001 mL)
A fresh urine sample

TEST PROTOCOL

1. Holding the loop *vertically*, immerse it in the urine sample. Then carefully withdraw it to obtain the correct volume of urine.

2. Inoculate the blood agar by making a single streak across the diameter of the plate.

3. Turn the plate 90° and, without flaming the loop, streak the urine across the entire surface of the agar as shown in Figure 7-6.

4. Again turn the plate 90° and streak a third time to assure maximum dispersal of any organisms present.

Photographic Atlas Reference Page 88

5. Invert, label and incubate the plate for 24 hours at 37°C.

6. Remove the plate from the incubator and count the resulting colonies. Also note any differing colony morphologies which would suggest possible colonization by more than one species.

7. Multiply the number of colonies by 100 if using a 0.01 mL loop; multiply by 1000 if using a 0.001 mL loop. (The number of cells contained in one milliliter is 100 times the number contained in 1/100th of a milliliter. Therefore, multiplying by the *reciprocal* of the volume gives the density in milliliters.)

8. Enter the *cell density* in mL in the table below.

PRECAUTION

⚠ Significant variability can be avoided by holding the loop in a vertical position while transferring.

REFERENCES

Baron, Ellen Jo, Lance R. Peterson and Sydney M. Finegold. 1994. Page 255 in *Bailey & Scott's Diagnostic Microbiology, 9th Ed.* Mosby–Year Book, Inc. St. Louis, Missouri.

Koneman, Elmer W., *et al.* 1997. Page 94 in *Color Atlas and Textbook of Diagnostic Microbiology, 5th Ed.* Lippincott-Raven Publishers, Philadelphia, PA.

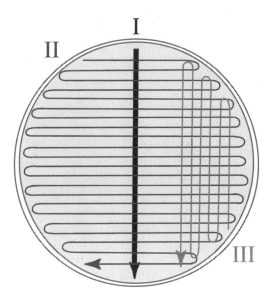

FIGURE 7-6

Semiquantitative streak method. Streak 1 is a simple streak line across the diameter of the plate. Streak 2 is a multiple streak at right angles to the first streak. As an alternate method, a third multiple streak may be added at right angles to the second streak.

URINE SAMPLE	COLONIES COUNTED (A)	RECIPROCAL OF THE VOLUME (100 OR 1000) (B)	ORIGINAL CELL DENSITY (A × B)

Bacterial Growth in a Closed System

MATERIALS

Two water baths set at 20°C and 37°C (It is preferable that these be shaker baths.)

One sterile 500 mL side-arm flask containing 49 mL of sterile tryptic soy broth (TSB)

One sterile 500 mL side-arm flask containing 50 mL of sterile tryptic broth (TSB) as a control

Spectrophotometer

12 hour culture of *Escherichia coli* in tryptic soy broth

Lab tissues

One sterile 2 mL pipette

Mechanical pipettor

Bunsen burner

[**NOTE TO INSTRUCTOR:** If your lab is shorter than 4 hours, modify this protocol as suggested in the Precautions section.]

TEST PROTOCOL

Twelve hours prior to your laboratory, a culture of *Escherichia coli* will be started for your class. This "starter culture" will be used by all groups in your lab for the experiment. Your instructor will assign you to either the "cool incubation" group (using the 20°C water bath) or the "warm incubation" group (using the 37°C water bath).

1. At the beginning of your laboratory session, collect the necessary materials. You will need a spectrophotometer, sterile 2 mL pipette, mechanical pipettor, the side-arm flasks containing 49 mL and 50 mL of broth, and lab tissues. Label the 50 mL flask "control."

2. Turn on the spectrophotometer and let it warm up.

3. Set the spectrophotometer's wavelength to 540 nm.

4. Use the absorbance (not % transmittance) scale for all readings.

5. When the spectrophotometer is warm, calibrate it.

   a. Use the left knob to set the left end of the scale to infinite absorbance. (On newer machines, this step is not necessary.)

   b. Wipe the side-arm of the control (50 mL TSB) flask with a tissue to remove finger prints. Then carefully place its side-arm in the sample port of the spectrophotometer.

   c. Use the right knob to adjust the setting to zero absorbance at the right end of the scale.

6. Label your growth flask with your group number or name using a piece of tape and permanent marker.

7. When you are ready to start, aseptically add 1.0 mL of the starter culture to the flask already containing 49 mL of broth. Mix it and immediately take a turbidity reading. This time is T_0. The reading is made by wiping the side-arm with tissue, carefully putting it into the sample port of the spectrophotometer, and reading the dial. Record the absorbance in the table below for time T_0.

8. Remove the growth flask carefully, then place it in the appropriate shaker bath for your group. Turn off the shaker when you do so and be careful not to break your flask or any others in the bath. Also, check the temperature to make sure it is correct.

9. Every 15 minutes after inoculating your flask, turn off the shaker bath, retrieve your flask and take a turbidity reading. You will need to clean the side-arm and calibrate the spectrophotometer each time. When finished, record your result in the table below and put your flask back into the shaker bath. You will take your last reading at T_{240}, that is, 4 hours after you started. Also collect data from another group that is incubating at the other temperature and record it in the table.

10. When you are done, plot your and the other lab group's absorbance values versus time. It is best to enter your data into a computer spreadsheet program and let it produce the graph, but if that is not possible, plot it by hand using the graph paper on the following page. From your graph, determine the length of time spent in any growth stages seen and the absorbance value at maximum stationary for each sample. Enter these numbers in the table provided

INCUBATION TEMPERATURE	T_0	T_{15}	T_{30}	T_{45}	T_{60}	T_{75}	T_{90}	T_{105}	T_{120}	T_{135}	T_{150}	T_{165}	T_{180}	T_{195}	T_{210}	T_{225}

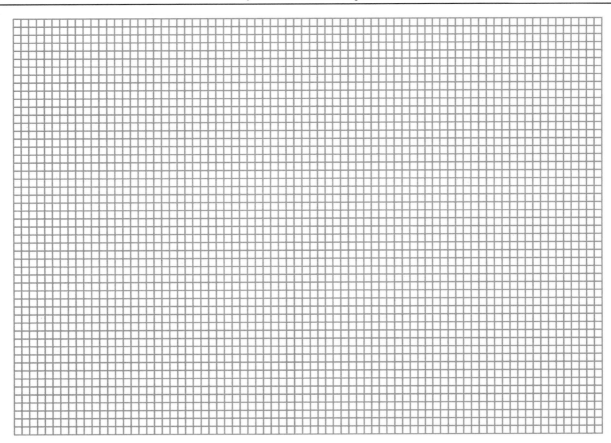

below. You will also calculate the mean growth rate constants and the mean generation times in exponential growth in steps #11 and #12.

11. To calculate the mean growth rate constant during exponential growth, choose two absorbance values in the linear part of the curve. The first (and smaller) value is A_1; the second (and larger) value is A_2. Determine the time interval in minutes (t) between A_1 and A_2. Substitute them into the equation to get the *mean growth rate constant*, k (shown below). Enter your calculated value for k in the table provided.

$$k = \frac{A_2 - A_1}{0.301t}$$

NOTE: The A_1 and A_2 terms are absorbances, which are already logarithms. If actual cell numbers had been obtained, then the numerator would be: $\log N_2 - \log N_1$.

Also, since bacterial reproduction typically occurs by binary fission, the logarithm of 2 (0.301) is used in the denominator. If a different reproductive pattern is used by the organism, then this term must be changed to reflect that difference.

12. The term k gives you the number of generations per minute. More useful, though, is the minutes per generation, or *mean generation time, g*. To calculate this, simply take the inverse of k. Enter your calculated value for g in the table provided.

$$g = \frac{1}{k}$$

13. This table is to record the features of your's and the other lab group's growth data as determined in Steps 10, 11 and 12. How do the values compare? Explain any differences.

INCUBATION TEMPERATURE	LAG PHASE (MINUTES)	EXPONENTIAL PHASE (MINUTES)	STATIONARY PHASE (MINUTES)	LAG PHASE ABSORBANCE	STATIONARY PHASE ABSORBANCE	k	g

14. This lab exercise may be modified to examine the effect of different variables. If your lab is equipped to do so, your instructor may assign or allow you to choose one of the following exercise modifications.

   ■ The same protocol may be used, but with different groups using different organisms.

   ■ The same protocol may be used, but with different groups using different incubation temperatures.

   ■ Different groups may use different volumes of initial inoculum. That is, instead of putting 1 mL into 49 mL of broth, use some other volume. Just be sure to adjust the amount of sterile broth so the total volume in the flask stays at 50 mL. Compare the same parameters as in the basic laboratory in your analysis.

   ■ Different groups may use starter cultures of different ages. Overnight cultures of 20, 15 and 10 hours work well. Follow the protocol as described (unless your lab is shorter than 4 hours). This experiment allows a class to compare the effects of the starter culture age on the subculture in the growth flask.

   ■ In all cases, compare the same parameters as in the basic laboratory in your analysis.

PRECAUTIONS

⚠ If your lab is shorter than the 4 hours this lab requires, use a larger starting inoculum: 2 mL for a 3.5 hour lab and 4 mL for a 3 hour lab. Be sure to adjust the volume of tryptic soy broth in the flask so the total volume equals 50 mL.

⚠ When reading the spectrophotometer's dial, view the needle so it obscures its image in the mirror to assure that you are observing from directly in front.

⚠ Place side-armed flasks into and remove them from the spectrophotometer in a straight line. If you push or pull at an angle, you risk breaking the side-arm and contaminating yourself, the table and the machine.

⚠ When taking readings, the flasks move a little. Set them in the same position each time and don't worry about the movement. Its effect is really negligible.

⚠ It is best to cover the opening into the spectrophotometer when taking readings to eliminate extra light. Use your hands (without moving the flask) or a special cover may be supplied by your instructor.

⚠ Use care when putting the flasks into or removing them from the shaker bath.

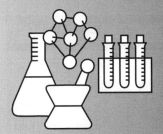

Medical, Food and Environmental Microbiology

Medical, food and environmental microbiology can be called the "real world" application of microbiological principles. In fact, many of the procedures taught in this manual and principles discussed in the *Photographic Atlas* are applied by medical personnel and governmental regulatory agencies on a daily basis.

In preparation for the exercises contained in this section, study the theoretical material and photographs in Section Seven of the *Photographic Atlas* or your text. It is important to understand the theory underlying the exercises in advance of actually performing them. The laboratory period can then be spent simply reinforcing concepts already learned. Read through each exercise and discuss it with your lab partners in time to ask your instructor for help if there are aspects you don't understand.

PRODUCING AND DETECTING MUTATIONS

In this unit you will perform exercises dealing with mutations, that is, alterations of the genetic material, DNA. The first exercise, which examines the effects of UV radiation on bacteria, illustrates some characteristics of a particular *mutagen* — how it causes damage, and how the bacteria are able to repair that damage. The second exercise — the "Ames Test" — illustrates a simple method of screening substances (commercial products) for mutagenicity, and in turn, carcinogenicity! As such, it is an important test in the fight against cancer.

Ultraviolet Radiation: Its Characteristics and Effects on Bacterial Cells

Exercise 8–1

MATERIALS

Four nutrient agar plates
Paper or cardboard masks with 1" to 2" cutouts
UV light (shielded for eye protection)
Sterile cotton applicators
24 hr broth culture of *Serratia marcescens*

TEST PROTOCOL

1. Using a sterile cotton applicator, streak inoculate a nutrient agar plate to form a bacterial lawn over the entire surface. Repeat this process for all remaining plates.

2. Number the plates 1, 2, 3 and 4.

3. Remove the lid from plate 1 and set it on a disinfectant-soaked towel. Place the plate under the UV light and cover it with a mask (Figure 8-1).

4. Turn the UV light on. After 30 seconds, turn the UV light off, remove the mask and plate, and replace the cover.

5. Repeat the process for plate 2.

6. Repeat the process for plate 3, but leave the UV light on for 3 minutes.

7. Finally, irradiate plate 4 the same as plate 3, but with the lid on.

8. Incubate plates 1, 3 and 4 for 24 to 48 hours at room temperature in an inverted position where they can receive natural light (*e.g.*, window sill).

9. Incubate plate 2 in the dark for the same time period.

10. Remove the plates and examine the growth patterns for evidence of light and dark repair.

PRECAUTIONS

⚠ Do not look at the UV light source while it is on.

⚠ The streaking should produce confluent growth on the agar surface (a "bacterial lawn"). The denser the lawn, the easier it is to interpret the results. So, make your streak lines close together.

⚠ If instructed to do so, your class may alter variables in this exercise. For instance: change the exposure times, change which plate gets the lid, *etc.*

REFERENCE

Lewin, Benjamin. 1990. *Genes IV*. Oxford University Press, Cambridge, Mass.

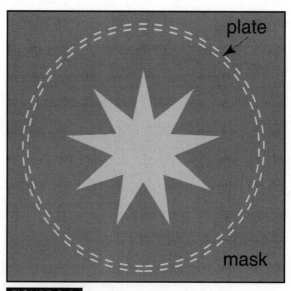

FIGURE 8-1

An example of a cardboard cutout placed over a Petri dish (shown as a shadow). The cutout may be any shape, but should leave the outer 25% of the plate masked.

Exercise 8–2

Ames Test

MATERIALS

Four Minimal Medium (MM) plates
Four Complete Medium (CM) plates
Two tubes containing 10.0 mL nutrient broth + 0.5% NaCl
Two bottles containing 90.0 mL nutrient broth + 0.5% NaCl
Centrifuge
Sterile centrifuge tubes
Small beaker containing alcohol and forceps
Bottle of 1x Vogel-Bonner salts (To make 1x Vogel-Bonner salts, add 1.0 mL 50x Vogel-Bonner solution to 49 mL water.) (Appendix B)
Sterile filter discs made with a paper punch
Two sterile Petri dishes (for soaking filter paper discs)
Sterile 10 mL pipettes
Container for disposal of supernatant (to be autoclaved)
DMSO
Test substance (any substance which has possible mutagenic properties and does not contain histidine or protein)
Latex gloves
Broth culture of *Salmonella typhimurium TA 1535**
Broth culture of *Salmonella typhimurium TA 1538**

TEST PROTOCOL

Day One

1. Soak four filter paper discs in DMSO and four filter paper discs in the test substance. You should wear latex gloves while handling the test substance.
2. Pipette 10.0 mL TA 1535 into a sterile centrifuge tube. Do the same with TA 1538, then label the tubes.

*** The cultures used for this exercise must be prepared as follows:**

1. 24 hours before the test, inoculate the two 10.0 mL broth tubes with TA 1535 and TA 1538 and incubate at 35°C together with the two sterile 90.0 mL broths.
2. 5½ hours before the test, pour the TA 1538 culture into one of the sterile 90.0 mL broths and return it to the incubator until time for the test.
3. 4 hours before the test, pour the TA 1535 culture into the other 90.0 mL sterile broth and return it to the incubator until time for the test.

3. Centrifuge the tubes on high speed for 10 minutes.

Photographic Atlas Reference Page 89

4. Being careful not to disturb the cell pellet at the bottom, decant the supernatant from each centrifuge tube (Figure 8-2).

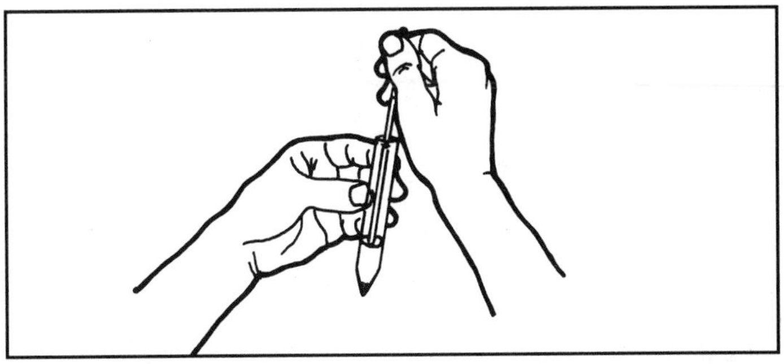

FIGURE 8-2 Decanting
This variation of decanting uses a Pasteur pipette and bulb to draw the supernatant away from the pellet. Discharge the air from the bulb before placing it in the tube, then slowly lower the pipette into the fluid and draw off the supernatant. Repeat this procedure until all of the supernatant has been removed. A little bit of the pellet may also be removed, but this can be minimized by working slowly to avoid agitation of the fluid and cells. Dispose of the supernatant in a container to be autoclaved.

5. Resuspend the cell pellets by adding 1.0 mL sterile 1x Vogel-Bonner salts to each tube and mixing well.
6. Using spread plate technique, inoculate two MM plates and two CM plates with 0.1 mL of the resuspended TA 1535. Do the same with TA 1538.
7. Sterilize the forceps by passing them through the Bunsen burner flame and allowing the alcohol to burn off.
8. Using the sterile forceps, place the discs in the centers of the plates as follows:

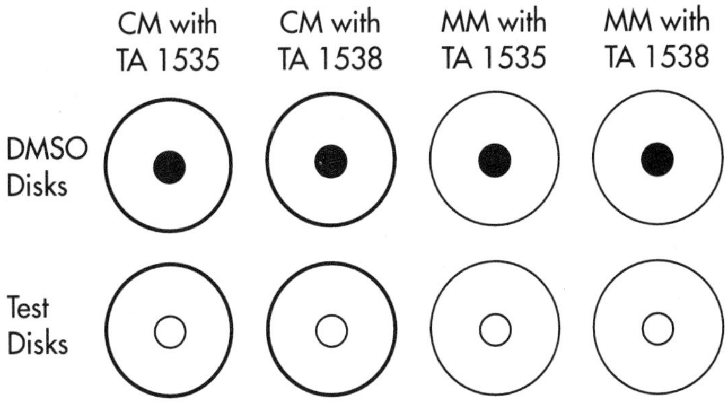

(Gently tap the disks down with the forceps to prevent them falling off when the plates are inverted.)

9. Incubate the plates aerobically at 35°C for 48 hours.

Day Two

1. Measure and compare the zones of inhibition on the CM plates (Figure 8-3). Count the colonies on the MM plates and compare (Figure 8-4) .

2. Record your results in the space provided.

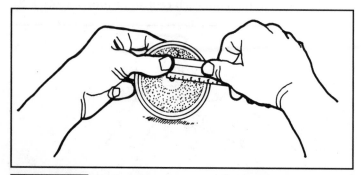

FIGURE 8-3 **Measuring the Zone of Inhibition**
Using a metric ruler, measure the shortest distance from the edge of the paper disc to the perimeter of the clearing. Do this with all the CM plates and record your measurement in mm.

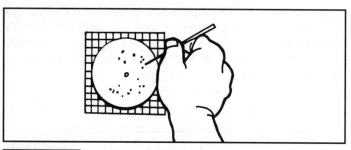

FIGURE 8-4 **Counting The Mutant Colonies**
Place the open plate on the colony counter, turn on the light and adjust the magnifying glass until all the colonies are visible. Using the grid in the background as a guide, count the larger colonies one section at a time. Mark each colony with a toothpick (to avoid counting it more than once) as you keep track of the number using a hand tally counter. Do this with all of the MM plates and record your results. Dispose of the toothpick and plates properly.

PRECAUTIONS

⚠ Be sure to load the centrifuge evenly and watch it at all times when it is running.

⚠ When reading the MM plates, count only the larger, fully grown colonies — these are the back-mutants. There will also be numerous punctiform colonies produced by auxotrophs that did not back-mutate to prototrophs and only grew until the histidine in the minimal medium was exhausted.

REFERENCES

Eisenstadt, Bruce C. Carlton, and Barbara J. Brown. 1994. Page 311 in *Methods for General and Molecular Bacteriology*, edited by Philipp Gerhardt, R. G. E. Murray, Willis A. Wood and Noel R. Krieg, American Society for Microbiology, Washington, DC.

Maron, D. M. and B. N. Ames. 1983. *Mutation Research*, 113:173-215.

	ZONE DIAMETER		COLONIES COUNTED	
	CM with TA 1535	CM with TA 1538	MM with TA 1535	MM with TA 1538
DMSO				
Test Substance				

DETERMINING THE ANTIBIOTIC OF CHOICE FOR TREATMENT

Antibiotic treatment of bacterial infection is often not as simple as identifying the organism, then going to the "book" to look up the proper medication. Due to increasing antibiotic-resistant (and *multiple antibiotic resistant!*) strains, the physician needs more specific information about the particular strain in question. If the generic treatment is ineffective, then the Kirby-Bauer test is called for. In this test, an isolated pathogen is incubated on a plate with a variety of antibiotic impregnated disks to determine which compound is most likely to resolve the patient's infection.

Exercise 8–3 — Antibiotic Sensitivity — Kirby-Bauer Test

MATERIALS

Four Mueller-Hinton agar plates

Antibiotic disks (commercially available): streptomycin, tetracycline, penicillin, chloramphenicol, sulfisoxazole and trimethoprim.

Antibiotic disk dispenser (See the dispenser in Figure 8-5. If you don't have one, you can use the forceps and alcohol listed below.)

Sterile cotton swabs

Metric ruler

Zone diameter interpretive table published by the National Committee for Clinical Laboratory Standards (NCCLS). Refer to page 94 of the *Photographic Atlas* for more information.

Small beaker with alcohol and forceps

Sterile saline

Sterile Pasteur pipettes

McFarland 0.5 standard with card (Appendix B)

Recommended organisms (grown in broth):
  Escherichia coli
  Staphylococcus aureus

TEST PROTOCOL

Day One

1. Gently agitate a broth culture tube and the McFarland standard until they reach their maximum turbidity.

2. Holding the broth and McFarland standard upright on the table in front of you, place the card behind them so that you can see the black line through the liquid in the tubes. As you can see, the line becomes distorted by the turbidity in the tubes (see Figure 7-8 in the *Photographic Atlas*). Use the black line to compare the level of turbidity of the two tubes. Dilute the broth with the sterile saline until it appears to have the same level of turbidity as the standard.

3. Repeat this process with the other broth.

4. Dip a sterile swab into one of the broths and wipe off the excess on the inside of the tube.

> **Photographic Atlas Reference Page 93**

5. Inoculate a Mueller-Hinton plate by streaking the entire surface of the agar three times with the swab (two at 90° angles and a third at a 45° angle). Inoculate two plates with each culture being tested. Be sure to use a different swab for each inoculation.

6. Label the plates with the organisms' names, your name and the date.

7. Apply the streptomycin, tetracycline, penicillin and chloramphenicol discs to the agar surface of one plate for each organism. You can apply the disks either singly using sterile forceps, or with a dispenser (Figure 8-5). The forceps can be sterilized by passing them through the Bunsen burner flame and allowing the

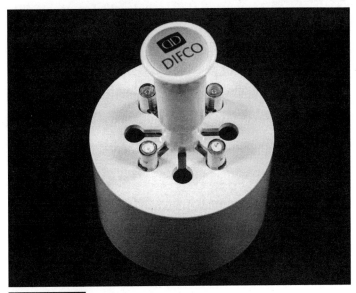

FIGURE 8-5   **Dispensing the Antibiotic Discs**
Center the dispenser over the open plate and press firmly only once to deposit the disks. Gently press each disk with sterile forceps to make sure it is making good contact with the agar surface. Cover the plate, invert it and place it in the incubator.

alcohol to burn off. Be sure to space the disks sufficiently (4 to 5 cm) to prevent overlapping zones of inhibition.

8. Press each disc gently with sterile forceps so that it makes good contact with the agar surface.

9. Using sterile forceps (as in #7 above) apply the sulfisoxazole and trimethoprim discs to the agar surface of the two remaining plates in the following manner:

   a. On the *E. coli* plate place the disks exactly 25 mm apart.

   b. On the *S. aureus* plate place the disks exactly 27 mm apart.

10. Invert the plates and incubate them aerobically at 35°C for 18 to 24 hours.

Day Two

1. Remove the plates from the incubator and measure the zones of inhibition (see Figure 8-6).

2. Record your results in the space provided.

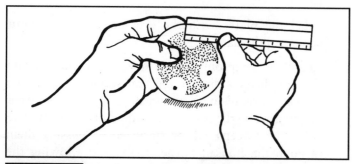

FIGURE 8-6 **Measuring the Antimicrobial Susceptibility Zones**
Place the inverted plate on the colony counter and, using a metric ruler, measure the entire diameter of each clearing. Record your results and dispose of the plates properly.

3. Observe the sulfisoxazole and trimethoprim plates for synergism (See Figure 7-11 in the *Photographic Atlas.*)

PRECAUTION

⚠ Take care to place the trimethoprim and sulfisoxazole disks exactly the recommended distances apart.

REFERENCES

Collins, C. H., Patricia M. Lyne, J. M. Grange. 1995. Page 128 in *Collins and Lyne's Microbiological Methods, 7th Ed.* Butterworth-Heinemann, UK.

Power, David A. and Peggy J. McCuen. 1988. Page 204 in *Manual of BBL® Products and Laboratory Procedures, 6th Ed.* Becton Dickinson Microbiology Systems, Cockeysville, MD.

Woods, Gail L. and John A. Washington. 1995. Page 1337 in *Manual of Clinical. Microbiology, 6th Ed.,* edited by Patrick R. Murray, Ellen Jo Baron, Michael A. Pfaller, Fred C. Tenover and Robert H. Yolken, ASM Press, Washington, DC.

Jorgensen, James H. et. al. 1994. *Performance Standards for Antimicrobial Susceptibility Testing; Fifth Informational Supplement, Vol. 14 no. 16,* NCCLS, Villanova, PA.

ORGANISM	ZONE DIAMETER			
	Streptomycin	Tetracylcine	Penicillin	Choloramphenicol

TESTS IDENTIFYING BACTERIAL CONTAMINATION IN SAMPLES

This unit consists of tests that qualitatively and/or quantitatively identify the presence of undesired bacteria in a sample. The first two tests — the membrane filter technique and the most probable number technique — are used to measure fecal contamination of water. The methylene blue reductase test is used to detect bacterial contamination of milk. Other foods may be checked using standard dilutions and the plate count method (Section Seven). Lastly, the Snyder test is a means of presumptively identifying the presence of decay-causing bacteria in the mouth, and thus, susceptibility to dental caries.

Membrane Filter Technique

MATERIALS

One Levine EMB plate

One sterile membrane filter (pore size 0.45 μm)

Sterile membrane filter suction apparatus (Figure 8-7)

100 mL water sample taken from any source where potability may be questionable (**Note:** Wear latex gloves while collecting and wipe the outside of the bottle with disinfectant afterward. Dispose of the gloves properly.)

Small beaker containing alcohol and forceps

Vacuum source (This can be either a vacuum pump or an aspirator connected to a faucet. Make sure that any faucet used is equipped with a properly installed anti-siphon device.)

TEST PROTOCOL

Day One

1. Sterilize the forceps by placing them in the Bunsen burner flame long enough to ignite the alcohol. Once the forceps are sterile, use them to place the filter (grid facing up) between the two halves of the filter housing. Clamp all parts together.

2. Insert the filter housing into the suction flask as shown in Figures 8-7.

3. Pour the water sample into the filter housing funnel and begin the suction.

4. When the water has passed through the membrane filter, stop the suction.

5. Sterilize the forceps again by flaming and carefully remove the filter.

6. Place the filter on the EMB agar being careful not to fold it or create air pockets under it (Figure 8-8).

7. Invert the plate and incubate it aerobically at 35°C for 24 hours.

Day Two

1. Remove the plate and count the colonies that are dark purple, have a black center or produce a green metallic sheen. If there are none, incubate the plate another 24 hours before reporting it as negative.

Photographic Atlas Reference Page 96

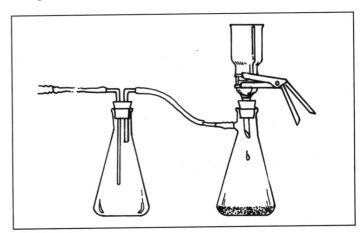

FIGURE 8-7   **Membrane Filter Apparatus**
Assemble the membrane filter apparatus as shown in this photograph. It is important to use two suction flasks (as shown) to avoid getting water into the vacuum source. Secure the flasks on the table as the tubing will make them top heavy.

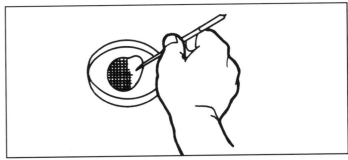

FIGURE 8-8   **Placing the Filter On the Agar Plate**
Using sterile forceps carefully place the filter onto the agar surface with the grid facing up. Try not to allow any air pockets under the filter since contact with the agar surface is essential for bacterial growth. Allow a few minutes for the filter to adhere to the agar before inverting the plate.

2. Determine the water sample's potability as follows:

$$\text{colonies per 100 mL} = \frac{\text{coliform colonies counted} \times 100}{\text{mL of sample filtered}}$$

PRECAUTION

⚠ Place the vacuum flasks in test tube baskets or otherwise secure them to the table to prevent the tubing from tipping them over.

REFERENCES

Chan, E. C. S., Pelczar, Jr., Krieg Noel R. 1986. Page 291 in *Laboratory Exercises In Microbiology*. McGraw-Hill Book Company.

Collins, C. H., Patricia M. Lyne, J. M. Grange. 1995. Page 270 in *Collins and Lyne's Microbiological Methods, 7th Ed.* Butterworth-Heinemann, UK.

DIFCO Laboratories. 1984. Page 515 in *DIFCO Manual, 10th Ed.* DIFCO Laboratories, Detroit, MI.

Mulvany, J. G. 1969. Page 205 in *Methods in Microbiology, Vol. 1,* edited by J. R. Norris and D. W. Ribbons, Academic Press Inc., New York.

Power, David A. and Peggy J. McCuen. 1988. Page 153 in *Manual of BBL® Products and Laboratory Procedures, 6th Ed.* Becton Dickinson Microbiology Systems, Cockeysville, MD.

MPN (Most Probable Number) Method for Total Coliform Determination

The completion of this exercise requires an understanding of serial dilutions, dilution factors and their calculations. Read and understand "MPN" in the *Photographic Atlas* before proceeding. For additional information refer to "Viable Count" in Section Seven.

MATERIALS

Fifteen lauryl tryptose broth (LTB) tubes (containing 10 mL broth and an inverted Durham tube)

Fifteen brilliant green lactose bile (BGLB) broth tubes (containing 10 mL broth and an inverted Durham tube). The number of tubes actually required will be determined by the results of the presumptive (LTB) phase.

Fifteen EC (*E. coli*) broth tubes (containing 10 mL broth and an inverted Durham tube). The number of tubes actually required will be determined by the results of the BGLB test.

Water sample suspected of fecal contamination (or unknown sample provided by your instructor)

Three 9.0 mL dilution tubes

Sterile 1.0 mL pipettes and pipettor

Water bath set at 45.5°C

Test tube rack

Labeling tape

TEST PROTOCOL

Day One

1. Arrange the 15 LTB tubes into three groups of five in the test tube rack.

2. Clearly label the groups 10^0, 10^{-1} and 10^{-2}.*

3. Aseptically transfer one mL of the water sample to each LTB tube in the group labeled "10^0."

4. Make a 10^{-1} dilution by adding 1.0 mL of the water sample to one of the 9.0 mL dilution tubes.

5. Add 1.0 mL of the 10^{-1} dilution to each of the LTB tubes so labeled.

6. Make a 10^{-2} dilution by adding 1.0 mL of the 10^{-1} dilution to one of the 9.0 mL dilution tubes.

7. Add 1.0 mL of the 10^{-2} dilution to each of the LTB tubes in the "10^{-2}" group.

8. Incubate the LTB tubes at 37°C for 48 hours.

Photographic Atlas Reference Page 97

Day Two

1. Remove the broths from the incubator and examine the Durham tubes for gas bubbles.

2. Using an inoculating loop or other transfer device, inoculate one BGLB broth with each LTB tube showing evidence of gas production. Make sure *each* BGLB tube is clearly labeled with the dilution factor of the LTB tube from which it is to be inoculated.

3. Inoculate EC broths with the positive LTB tubes in the same manner as the BGLB above. Again be sure to *clearly* label all EC tubes with the original LTB dilution factor.

4. Incubate the BGLB at 37°C for 48 hours. Incubate the EC tubes in the 45.5°C water bath for 48 hours.

Day Three

1. Remove all tubes from the incubator and water bath and examine the Durham tubes for gas bubbles. Count the positive BGLB tubes and enter your results in the table at the top of page 132. Be sure to enter the numbers as a fraction with the number of positive tubes as the numerator and the number of tubes *in the group* as the denominator (see example). All of these numbers will be necessary for calculation of the MPN if your combination of positives is other than those listed in Table 8.1.

Example:

Dilution Factor	10^0	10^{-1}	10^{-2}
# Positive BGLB Tubes	5/5	3/5	1/5

In this example the combination of positives is 5–3–1.

2. Determine the total coliform MPN from Table 8.1 by using your "combination of positives" from the BGLB test (*i.e.*, three numerators).

3. Count the positive EC results in the same manner as the BGLB test. Record the results in the table at the top of page 132.

4. Determine the *E. coli* MPN using Table 8.1.

---

* The dilution factors 10^0, 10^{-1} and 10^{-2} represent the portion of water transferred to a tube that is *original sample*. For example, $10^{-1} = 0.1$, therefore a broth labeled 10^{-1} which received 1.0 mL of diluted water sample actually receive only 0.1 mL of the original sample and 0.9 mL of diluent.

Dilution Factor	10^0	10^{-1}	10^{-2}
Positive BGLB Tubes			
Positive EC Tubes			

5. Enter your results below. (Final results should be recorded as MPN/100 mL.)

Total coliform MPN/100 mL	
E. coli MPN/100 mL	

6. If your combination of positives differs from those listed in Table 8.1, you can calculate MPN using Thomas' simple formula (the MPN values derived using this formula are estimates and will vary slightly from the values in the table):

$$\text{MPN/100 mL} = \frac{\text{number of positive tubes} \times 100}{\sqrt{\text{mL sample in negative tubes} \times \text{mL sample in all tubes}}}$$

7. Calculation of the 5–3–1 combination in the example on page 131 would be as follows:

$$\text{MPN/100 mL} = \frac{\text{number of positive tubes} \times 100}{\sqrt{\text{mL sample in negative tubes} \times \text{mL sample in all tubes}}}$$

$$\text{MPN/100 mL} =$$

$$\frac{9 \times 100}{\sqrt{(.24 \times 5.55)}} = \frac{900}{\sqrt{1.332}} = \frac{900}{1.154} = 780$$

REFERENCE

Greenberg, Arnold E., *et al.* 1992. *Standard Methods for the Examination of Water and Wastewater, 18th Ed.* American Public Health Association, Washington, DC.

TABLE 8.1 MPN index and 95% confidence limits for various combinations of positive results.

Combination of Positives	MPN Index/ 100 mL	95% Confidence Limits Lower	95% Confidence Limits Upper	Combination of Positives	MPN Index/ 100 mL	95% Confidence Limits Lower	95% Confidence Limits Upper
0–0–0	<20	—	—	4–3–0	270	120	670
0–0–1	20	10	100	4–3–1	330	150	770
0–1–0	20	10	100	4–4–0	340	160	800
0–2–0	40	10	130	5–0–0	230	90	860
1–0–0	20	10	110	5–0–1	300	100	1100
1–0–1	40	10	150	5–0–2	400	200	1400
1–1–0	40	10	150	5–1–0	300	100	1200
1–1–1	60	20	180	5–1–1	500	200	1500
1–2–0	60	20	180	5–1–2	600	300	1800
2–0–0	40	10	170	5–2–0	500	200	1700
2–0–1	70	20	200	5–2–1	700	300	2100
2–1–0	70	20	210	5–2–2	900	400	2500
2–1–1	90	30	240	5–3–0	800	300	2500
2–2–0	90	30	250	5–3–1	1100	400	3000
2–3–0	120	50	290	5–3–2	1400	600	3600
3–0–0	80	30	240	5–3–3	1700	800	4100
3–0–1	110	40	290	5–4–0	1300	500	3900
3–1–0	110	40	290	5–4–1	1700	700	4800
3–1–1	140	60	350	5–4–2	2200	1000	5800
3–2–0	140	60	350	5–4–3	2800	1200	6900
3–2–1	170	70	400	5–4–4	3500	1600	8200
4–0–0	130	50	380	5–5–0	2400	1000	9400
4–0–1	170	70	450	5–5–1	3000	1000	13000
4–1–0	170	70	460	5–5–2	5000	2000	20000
4–1–1	210	90	550	5–5–3	9000	3000	29000
4–1–2	260	120	630	5–5–4	16000	6000	53000
4–2–0	220	90	560	5–5–5	≥16000	—	—
4–2–1	260	120	650				

Methylene Blue Reductase Test

MATERIALS

Three sterile test tubes each containing 10.0 mL of fresh milk, labeled "A," "B" and "Control"

Sterile 1 mL pipettes

Hot water bath set at 35°C

Methylene blue solution

Recommended organisms (on solid medium):
    Escherichia coli

Clock or wristwatch

TEST PROTOCOL

1. Inoculate tube A with *Escherichia coli*. (The addition of *E. coli* to the milk simulates milk of poor quality. However, any milk of questionable quality can be tested with this procedure in place of the "spiked" sample. To do this, skip step #1 and begin at step #2. Use as many samples as you like but be sure to label them clearly.)

2. Aseptically add 1.0 mL methylene blue solution to test tube A and to test tube B. Cap the tubes tightly and invert them several times to mix thoroughly.

3. Place tubes A and B in the hot water bath and note the time.

4. Place the control tube in the refrigerator.

5. After a 5 minute incubation remove the tubes, invert them *once* to mix again then return them to the water bath. Record the time in the table below under "STARTING TIME."

6. Using the control tube for color comparison, check tubes A and B at 30 minute intervals and record the time when each becomes white. Poor quality milk takes less than 2 hours; good quality milk takes longer than 6 hours.

*Photographic Atlas
Reference
Page 99*

7. Using the table below, calculate the time it takes for each milk sample to become white.

PRECAUTION

⚠ Be sure to record the *clock* times in the table under "STARTING TIME" and "ENDING TIME." The "ELAPSED TIME" is the difference between the starting and ending times and is the actual number of hours or minutes taken to complete the reaction.

REFERENCES

Bailey, R. W., and E. G. Scott. 1966. Page 114 and 306 in *Diagnostic Microbiology, 2nd Ed.* C. V. Mosby Company, St. Louis, MO.

Benathen, Isaiah. 1993. Page 132 in *Microbiology With Health Care Applications.* Star Publishing Company, Belmont, CA.

Power, David A. and Peggy J. McCuen. 1988. Page 62 in *Manual of BBL® Products and Laboratory Procedures, 6th Ed.* Becton Dickinson Microbiology Systems, Cockeysville, Md.

Richardson (ed.). 1985. *Standard Methods for the Examination of Dairy Products, 15th Ed.* American Public Health Association, Washington DC.

TUBE	STARTING TIME T_s (Milk is blue)	ENDING TIME T_e (Milk is white)	ELAPSED TIME ($T_e - T_s$)	MILK QUALITY
A				
B				

Exercise
8–7

Snyder Test

MATERIALS

Hot water bath set at 45–50°C
Small sterile beakers
Sterile 1 mL pipettes with bulbs
Two Snyder agar tubes

TEST PROTOCOL

1. Collect a small sample of saliva (about 0.5 mL) in the sterile beaker.

2. Aseptically add 0.2mL of the sample to a molten Snyder agar tube (from the water bath) and roll it between your hands until the saliva is uniformly distributed throughout the agar.

3. Allow the agar to cool to room temperature; do not slant.

4. Incubate with an uninoculated control at 35°C for up to 72 hours.

5. Check the tubes at 24 hour intervals for yellow color formation.

6. Record your results and determine your susceptibility to tooth decay using the information below and on page 100 of the *Photographic Atlas*.

Photographic Atlas
Reference
Page 100

REFERENCES

DIFCO Laboratories. 1984. Page 619 in *DIFCO Manual, 10th Ed.* DIFCO Laboratories, Detroit, MI.
Power, David A. and Peggy J. McCuen. 1988. Page 247 in *Manual of BBL® Products and Laboratory Procedures, 6th Ed.* Becton Dickinson Microbiology Systems, Cockeysville, MD.

TABLE OF RESULTS	
RESULT	**INTERPRETATION**
Yellow at 24 hours	High susceptibility to dental caries
Yellow at 48 hours	Moderate susceptibility to dental caries
Yellow at 72 hours	Slight susceptibility to dental caries
Yellow at >72 hours	Negative

TIME	24 HRS	48 HRS	72 HRS
COLOR			

Based on the results of this test, I have a _____ susceptibility to dental caries.

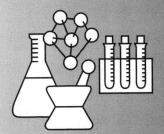

Hematology, Immunology, and Serology

*I*n this section you will be looking at cells and organs involved in the body's defenses, and at how the products of some of these cells (*i.e.*, "antibodies") are used *in vitro* as diagnostic tools. In the hematology and immunology units, you will have the opportunity to perform a differential blood cell count and examine various organs of the immune system. You will also perform tests used to detect the presence of specific antigens or antibodies in a sample.

HEMATOLOGY AND IMMUNOLOGY

The exercises in this unit deal with the blood and immune systems. In the first exercise, a differential blood cell count will be done. These are done by clinical laboratories as a means of diagnosing (or eliminating from consideration) certain pathological conditions. While differential counts are automated now, it is good training to perform one "the old-fashioned way" using a blood smear, and a microscope to get an idea of the principle behind the technique. The second exercise is a survey of immune organ structures to complement the theory presented in your lecture.

Exercise 9–1

Differential Blood Cell Count

MATERIALS

Commercially prepared human blood smear slides (Wright's or Giemsa stain)

Optional: Commercially prepared abnormal human blood smear slides (*e.g.*, infectious mononucleosis, eosinophilia, or neutrophilia)

TEST PROTOCOL

1. Obtain a blood smear slide and locate a field where the cells are spaced far enough apart to allow easy counting. (The cells should be fairly dense on the slide, but not overlapping.)

2. Using the oil immersion lens, scan the slide using the pattern shown in Figure 9-1.

3. Make a tally mark in the appropriate box of the table for the first 100 leukocytes you see.

4. Calculate percentages and compare your results with the accepted normal values.

5. Repeat with a pathological blood smear (if available).

PRECAUTIONS

> **Photographic Atlas Reference Page 101**

⚠ Remember that a microscope image is inverted. If you want the image to move left, you must move the slide to the right.

⚠ Be careful not to overlap fields when scanning the specimen. Choose a "landmark" blood cell at the right side of the field and move the slide horizontally until that cell disappears off the left side.

⚠ Avoid diagonal movement of the slide. As you scan, use the mechanical stage knobs separately to move the slide up and back or to the right in straight lines.

REFERENCES

Brown, Barbara A. 1993. *Hematology – Principles and Procedures, 6th Ed.* Lea and Febiger, Philadelphia, PA.

Eroschenko, Victor P. 1993. *di Fiore's Atlas of Histology with Functional Correlations, 7th Ed.* Lea and Febiger, Philadelphia, PA.

Junqueira, L. Carlos, Jose Carneiro and Robert O. Kelley. 1995. *Basic Histology, 8th Ed.* Appleton & Lange, Norwalk, CT.

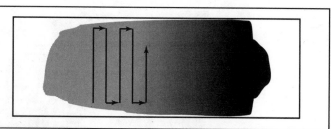

FIGURE 9-1 **Follow a Systematic Path**
A systematic scanning path is used to avoid wandering around the slide and perhaps counting some cells more than once.

NORMAL BLOOD						
	MONOCYTES	LYMPHOCYTES	SEGMENTED NEUTROPHILS	BAND NEUTROPHILS	EOSINOPHILS	BASOPHILS
Number						
Percentage						
Expected Percentage	3 – 7%	25 – 33%	55 – 65% (All Neutrophils)		1 – 3%	0.5 – 1%

ABNORMAL BLOOD						
	MONOCYTES	LYMPHOCYTES	SEGMENTED NEUTROPHILS	BAND NEUTROPHILS	EOSINOPHILS	BASOPHILS
Number						
Percentage						
Expected Percentage	3 – 7%	25 – 33%	55 – 65% (All Neutrophils)		1 – 3%	0.5 – 1%

Organs and Cells of the Immune System

MATERIALS

Prepared slides of:
  loose areolar tissue
  liver (stained for Kupffer cells)
  lymph node
  thymus
  tonsil
  spleen
  lung
  ileum

PROCEDURE

Examine the slides. On each, identify distinguishing characteristics and components relevant to the immune system.

1. loose areolar tissue: mast cells, macrophages, collagen and elastic fibers

2. liver: Kupffer cells, plates of hepatocytes, sinusoids, central veins

3. lymph node: lymph follicles (nodules) with germinal centers, lymphocytes, macrophages in sinuses (possibly), capsule

4. thymus: lobules, lymphocytes, thymic corpuscles, capsule

5. tonsil: crypts, lymph follicles (nodules), lymphocytes

**Photographic Atlas
Reference
Page 103**

6. spleen: capsule, lymph nodules (white pulp) with central artery, blood sinusoids (red pulp)

7. lung: lymph nodules lining respiratory tree, lymphocytes

8. ileum: Peyer's patch with lymph nodules, lymphocytes

PRECAUTION

⚠ Regions of lymphatic tissue are identifiable even at low power. Use higher magnification for examining detail.

REFERENCES

Eroschenko, Victor P. 1993. *di Fiore's Atlas of Histology with Functional Correlations, 7th Ed.* Lea and Febiger, Philadelphia, PA.
Junqueira, L. Carlos, Jose Carneiro and Robert O. Kelley. 1995. *Basic Histology, 8th Ed.* Appleton & Lange, Norwalk, CT.

SIMPLE SEROLOGICAL REACTIONS

Antigen-antibody reactions are very specific, and they will occur *in vitro* as well as *in vivo*. Serology is the discipline that exploits this specificity as an *in vitro* diagnostic tool.

Two simple serological reactions, agglutination and precipitation, are used in the following exercises because they result in the formation of complexes which can be viewed with the naked eye. While agglutination and precipitation largely have been replaced by more sensitive techniques in the clinical laboratory, they provide a simple and visual means of demonstrating the fundamental basis for all serological reactions — the binding of antigens to antibodies. (For more advanced techniques such as ELISA,

immunofluorescence, and Western Blot, please see Section 9 in the *Photographic Atlas*.)

Two examples of precipitation reactions are given: the precipitin ring test and the double gel radial immunodiffusion test. The former may be used to identify antigens or antibodies in a sample; the latter is used to compare antigens in more than one sample. The slide agglutination test can be an important diagnostic (and very specific) tool for the identification of organisms. It is especially useful for serotyping large genera such as *Salmonella*. Hemagglutination, which detects specific antigens on RBCs, is the standard test for determining blood type. Other hemagglutination tests are still used for diagnosis of infections.

Exercise 9–3   Precipitation Reactions — Precipitin Ring

MATERIALS

     One clean Durham tube
     Horse serum
     Horse antiserum
     Pasteur pipettes

TEST PROTOCOL

1. Carefully add horse antiserum to the Durham tube. Fill from the bottom of the tube until it is about ⅓ full.

2. Add the horse serum in such a way that a sharp and distinct second layer is formed without any mixing of the two solutions.

3. Incubate at 35°C for one hour.

4. Observe the tube for the characteristic ring formed at the equivalence zone. If after one hour there is no

ring, place the tube in the refrigerator for 12 to 24 hours and then recheck.

> *Photographic Atlas*
> *Reference*
> *Page 107*

PRECAUTION

⚠ It is critical that the serum be *very* carefully placed on top of the anti-serum *without any mixing* of the two solutions. Success can usually be achieved by allowing the serum to slowly trickle down the inside of the glass.

REFERENCE

Lam, Joseph S. and Lucy M. Mutharia. 1994. Page 120 in *Methods for General and Molecular Bacteriology*, edited by Philipp Gerhardt, R. G. E. Murray, Willis A. Wood and Noel R. Krieg, American Society for Microbiology, Washington, DC.

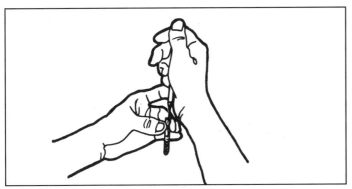

FIGURE 9-2   **Placing the Antibody In the Tube**
When adding the antibody to the tube, fill from the bottom to avoid creating air bubbles. Fill about ⅓ full.

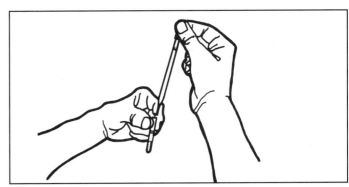

FIGURE 9-3   **Layering the Antigen On Top of the Antibody**
When adding the antigen layer to the tube containing antiserum, it is essential that there be no mixing of the two solutions. Place the pipette containing the antigen into the tube about ½ cm from the antiserum and let it slowly trickle down the inside of the glass. Allow to stand for one hour undisturbed.

Precipitation Reactions — Double-Gel Immunodiffusion

MATERIALS

One saline agar plate

3 mm punch (a glass dropper with a 3 mm diameter tip will work)

Template for cutting wells (Figure 9-4)

Anti-bovine albumin

Anti-horse albumin

10% bovine serum (Prepared by adding 9 drops physiological saline to 1 drop 100% serum)

10% horse serum (Prepared by adding 9 drops physiological saline to 1 drop 100% serum)

Disposable micropipettes

TEST PROTOCOL

1. Center the saline agar plate over the template in Figure 9-4.

2. Using the punch or glass dropper, cut the 7 wells in the agar as shown in Figure 9-5. If the small agar disks don't come out with the dropper, use suction

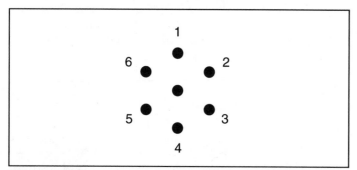

FIGURE 9-4 **Double-Gel Immunodiffusion Well Template**
Use this template as a guide when boring the wells in the saline agar plate.

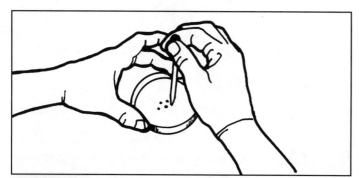

FIGURE 9-5 **Boring Wells In the Saline Agar Plate**
When cutting the wells in the agar, press straight down with the cutter; *do not twist.* Twisting the cutter or dropper will create fissures in the agar which could disrupt the diffusion of the sera.

with the dropper and bulb to dislodge and remove them.

Photographic Atlas Reference Page 107

3. Number the peripheral wells #1 through #6 as shown in Figure 9-4.

4. In a small test tube or mixing cup prepare a 1:1 mixture of anti-horse albumin and anti-bovine albumin.

5. Using a *different* micropipette for each transfer, carefully fill the wells as follows (Figure 9-6):

   a. Fill the center well with the anti-horse/anti-bovine mixture.

      1. Fill wells #1 and #2 with horse serum.
      2. Fill wells #3 and #4 with bovine serum.
      3. Fill well #5 with physiological saline.
      4. Fill well #6 with horse *and* bovine serum.

6. Cover the plate and allow it to sit undisturbed for 30 minutes.

7. Incubate at room temperature for up to 72 hours or until precipitation lines appear.

8. Examine the plate for precipitation patterns (Figure 9-7).

PRECAUTIONS

⚠ Avoid lifting the agar when removing the punched-out disks.

⚠ Be careful when adding the solutions not to overfill the wells.

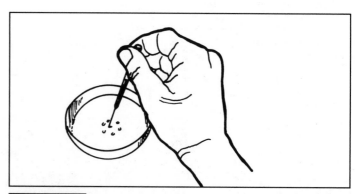

FIGURE 9-6 **Filling the Wells**
Place the tip of the pipette in the bottom of the well. Fill slowly to prevent creating air bubbles and to minimize spilling serum over the sides.

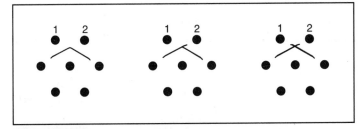

FIGURE 9-7   **Precipitation Patterns**

The diagram above shows three possible precipitation patterns formed between the antibody in the center well and antigens in wells #1 and #2. The pattern on the left demonstrates identity between the antigens. This indicates that the antigens are identical. The pattern in the middle demonstrates partial identity, which means that the antigens are related but not identical. The pattern on the right shows nonidentity; the antigens are not related. Refer also to Figure 9-5 in the *Photographic Atlas.*

REFERENCES

Lam, Joseph S. and Lucy M. Mutharia. 1994. Page 120 in *Methods for General and Molecular Bacteriology*, edited by Philipp Gerhardt, R. G. E. Murray, Willis A. Wood and Noel R. Krieg, American Society for Microbiology, Washington, DC.

Ouchterlony, O. 1968. Page 20 in *Handbook of Immunodiffusion and Immunoelectrophoresis*. Ann Arbor Science Publishers, Ann Arbor, MI.

Agglutination Reactions — Slide Agglutination

MATERIALS

Salmonella H antigen
Salmonella O antigen
Salmonella anti-H antiserum
One microscope slide
Toothpicks
Grease pencil

TEST PROTOCOL

1. Using a grease pencil, draw two circles approximately the size of a dime on a microscope slide.

2. Place a drop of *Salmonella* anti-H antiserum in each circle.

3. Place a drop of Salmonella O antigen in one circle and a drop of Salmonella H antigen in the other circle. Using a *different* toothpick for each circle, mix until each of the antigens is completely emulsified with the antiserum. Discard the toothpicks in a biohazard container.

4. Allow the slide to sit for a few minutes and observe for agglutination.

PRECAUTIONS

Photographic Atlas Reference Page 109

⚠ To make the exercise more realistic, the instructor may cover the labels on the two *Salmonella* antigen vials and have you determine which contains the H antigen.

⚠ Use a clean toothpick for each antigen to reduce the chance of false positives.

REFERENCES

Collins, C. H., Patricia M. Lyne, J. M. Grange. 1995. Page 118 in *Collins and Lyne's Microbiological Methods, 7th Ed.* Butterworth-Heinemann, UK.

Lam, Joseph S. and Lucy M. Mutharia. 1994. Page 120 in *Methods for General and Molecular Bacteriology*, edited by Philipp Gerhardt, R. G. E. Murray, Willis A. Wood and Noel R. Krieg, American Society for Microbiology, Washington, DC.

Agglutination Reactions — Blood Typing

MATERIALS

Blood typing anti-A antiserum
Blood typing anti-B antiserum
Blood typing anti-Rh antiserum
Two microscope slides
Toothpicks
Grease pencil
Sterile lancets
Alcohol wipes
Small adhesive bandages
Sharps container
Disposable latex gloves

TEST PROTOCOL

1. Draw two circles with your grease pencil on one microscope slide. Label one circle "A" and the other "B."

2. Draw a single circle with your grease pencil in the center of a second microscope slide. Label it "Rh."

3. Place a drop of anti-A antiserum in the "A" circle.

4. Place a drop of anti-B antiserum in the "B" circle.

5. On the second microscope slide, place a drop of anti-Rh antiserum.

6. Clean the tip of your finger with an alcohol wipe. Let the alcohol dry.

7. Open a lancet package and remove the lancet, being careful not to touch the tip before you use it.

8. Prick the end of your finger and immediately place a drop of blood beside each drop of antiserum. Do not touch the antisera with your finger. It's OK to have someone else prick your finger, but make sure he or she wears protective gloves.

9. Discard the lancet in the sharps container.

10. Using a circular motion, mix each set of drops with a toothpick. Be sure to use a *different toothpick* for each antiserum.

Photographic Atlas Reference Page 109

11. Gently rock the slides back and forth for a few minutes or until agglutination occurs.

12. After the agglutination reaction is complete, record the results in the table below.

ANTISERUM	AGGLUTINATION +/−
ANTI-A	
ANTI-B	
ANTI-Rh	

13. Using the information below and in Figures 9-8 and 9-9 in the *Photographic Atlas*, determine your blood type and enter it in the space provided in the chart on the following page.

PRECAUTIONS

⚠ Although not absolutely essential, a "light box" used to gently warm the slides while you tilt them improves the hemagglutination reaction.

⚠ Do not touch the antisera with your finger when placing blood on the slide.

TABLE OF RESULTS				
ANTI-A ANTISERUM	ANTI-B ANTISERUM	ANTI-Rh ANTISERUM	INTERPRETATION	SYMBOL
Agglutination	No Agglutination	Agglutination	A antigen present / Rh antigen present	A+
		No Agglutination	A antigen present / Rh antigen absent	A−
No agglutination	Agglutination	Agglutination	B antigen present / Rh antigen present	B+
		No agglutination	B antigen present / Rh antigen absent	B−
Agglutination	Agglutination	Agglutination	A and B antigens present / Rh antigen present	AB+
		No agglutination	A and B antigens present / Rh antigen absent	AB−
No Agglutination	No Agglutination	Agglutination	A and B antigens absent / Rh antigen present	O+
		No agglutination	A and B antigens absent / Rh antigen absent	O−

My blood type is:_____

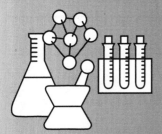

Eukaryotic Microbes

Microbiology in college courses is usually dominated by bacteriology, but the discipline also includes eukaryotic microscopic organisms, some of which are of medical importance. In this section, representative protozoans and the truly microscopic fungi are examined. These are followed by an exercise dealing with parasitic helminths (worms). While not typically microscopic, they often are encountered by microbiologists responsible for their identification as they examine various patient samples. For this reason, these parasites have entered the domain of microbiology.

Exercise
10-1

Fungi

MATERIALS

Photographic Atlas
Reference
Page 143

Agar slant of *Saccharomyces cerevisiae*
Plate culture of *Aspergillus spp.*
Plate culture of *Penicillium spp.*
Plate culture of *Rhizopus spp.*
Gram's iodine stain
Dissecting microscope
Prepared slides of:
　Aspergillus spp. conidiophore
　Candida albicans
　Penicillium spp. conidiophore
　Rhizopus spp. sporangia
　Rhizopus spp. gametangia

PROCEDURE

Yeasts

1. Make a wet mount slide of *Saccharomyces cerevisiae* and stain with iodine. Observe under high dry and oil immersion. Identify vegetative cells and budding cells. Sketch what you see in the space below.

2. Observe prepared slides of *Candida albicans*. Identify vegetative cells and budding cells. Sketch what you see in the space below.

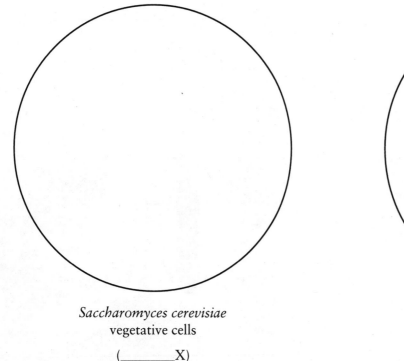

Saccharomyces cerevisiae
vegetative cells

(_____X)

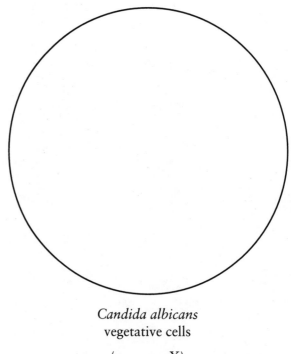

Candida albicans
vegetative cells

(_____X)

Molds

1. Observe the plate culture of *Rhizopus* using the dissecting microscope. Identify hyphae, rhizoids, and sporangia. Sketch and label what you see in the space below. DO NOT OPEN THE PLATE LID OR YOU WILL SPREAD SPORES AND CONTAMINATE THE LABORATORY.

2. Examine prepared slides of *Rhizopus* sporangia using medium and high dry powers. Identify the following: sporangiophores, sporangia, and spores. Sketch and label what you see in the space below.

3. Examine prepared slides of *Rhizopus* gametangia using medium and high dry power. Identify the following: progametangia, gametangia, young zygosporangia, mature zygosporangia. Sketch and label what you see in the space below.

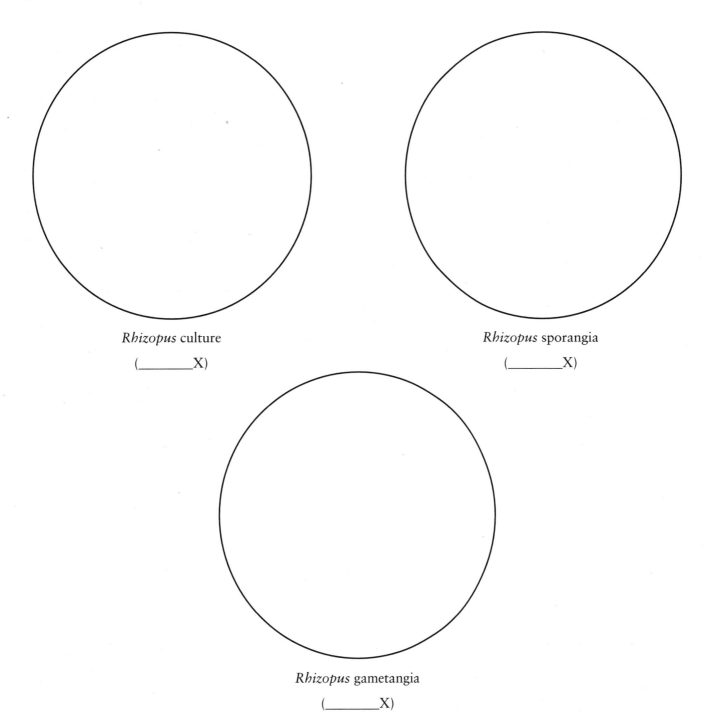

Rhizopus culture

(_____X)

Rhizopus sporangia

(_____X)

Rhizopus gametangia

(_____X)

4. Observe the plate culture of *Penicillium* using the dissecting microscope. Identify hyphae and ascospores. Sketch and label what you see in the space below. DO NOT REMOVE THE PLATE LID OR YOU WILL SPREAD SPORES AND CONTAMINATE THE LABORATORY.

5. Observe prepared slides of *Penicillium* conidiophores. Identify the following: hyphae, phialides, branched metulae, and chains of conidia. Sketch and label what you see in the space below.

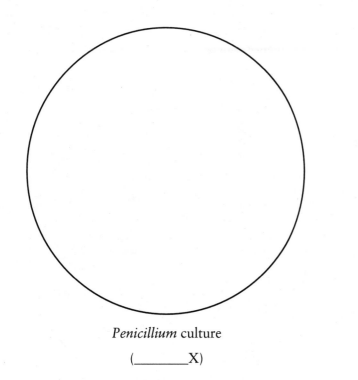

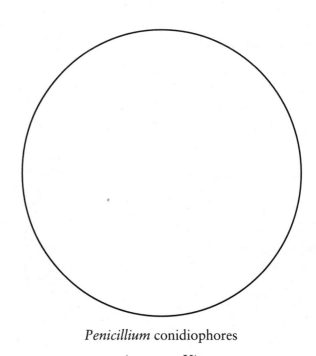

Penicillium culture

(_____X)

Penicillium conidiophores

(_____X)

6. Observe the plate culture of *Aspergillus* using the dissecting microscope. Identify hyphae and ascospores. Sketch and label what you see in the space below. DO NOT REMOVE THE PLATE LID OR YOU WILL SPREAD SPORES AND CONTAMINATE THE LABORATORY.

7. Observe prepared slides of *Aspergillus* conidiophores. Identify hyphae, conidiophores, phialides and conidia. Sketch and label what you see in the space below.

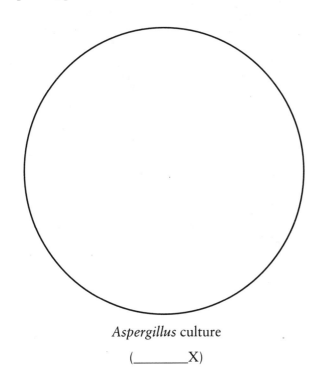

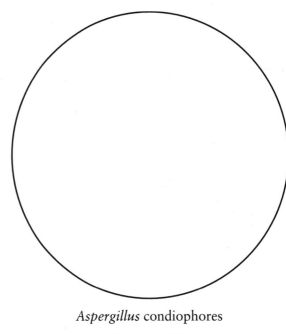

Aspergillus culture

(_____X)

Aspergillus condiophores

(_____X)

REFERENCES

Collins, C.H., Patricia M. Lyne and J.M. Grange. 1995. Chapter 51 in *Collins and Lyne's Microbiological Methods, 7th Ed.* Butterworth-Heineman, Oxford.

Hadley, W. Keith and Valerie L. Ng. 1995. Chapter 62 in *Manual of Clinical Microbiology, 6th Ed.*, edited by Patrick R. Murray, Ellen Jo Baron, Michael A. Pfaller, Fred C. Tenover, and Robert H. Yolken. American Society for Microbiology, Washington, D.C.

Kennedy, Michael J. and Lynne Sigler. 1995. Chapter 64 in *Manual of Clinical Microbiology, 6th Ed.*, edited by Patrick R. Murray, Ellen Jo Baron, Michael A. Pfaller, Fred C. Tenover, and Robert H. Yolken. American Society for Microbiology, Washington, D.C.

Koneman, Elmer W., Stephen D. Allen, William M. Janda, Paul C. Schreckenberger and Washington C. Winn, Jr. 1997. Chapter 19 in *Color Atlas and Textbook of Diagnostic Microbiology, 5th Ed.* J.B. Lippincott Company, Philadelphia, PA.

Mauseth, James D. 1995. Chapter 20 in Botany An Introduction to Plant Biology, 2nd Ed. Saunders College Publishing, Philadelphia, PA.

Raven, Peter H., Ray F. Evert and Susan Eichhorn. 1992. Chapter 12 in *Biology of Plants, 5th Ed.* Worth Publishers, New York, NY.

Warren, Nancy G. and Kevin C. Hazen. 1995. Chapter 61 in *Manual of Clinical Microbiology, 6th Ed.*, edited by Patrick R. Murray, Ellen Jo Baron, Michael A. Pfaller, Fred C. Tenover, and Robert H. Yolken. American Society for Microbiology, Washington, D.C.

Exercise 10–2

Protozoans

MATERIALS

Photographic Atlas
Reference
Page 151

Fresh culture of *Amoeba spp.*
Fresh culture of *Paramecium spp.*
Fresh culture of *Euglena spp.*
Methyl cellulose
Clean microscope slides and cover glasses
Methylene blue stain
Prepared slides of:
    Entamoeba histolytica trophozoite and cyst
    Entamoeba coli trophozoite and cyst
    Balantidium coli trophozoite and cyst
    Giardia lamblia trophozoite and cyst
    Trichomonas vaginalis trophozoite
    Leishmania donovani promastigote
    Trypanosoma spp.
    Plasmodium spp.
    Toxoplasma gondii trophozoite

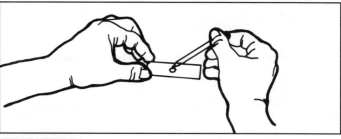

FIGURE 10-1   **Place a Drop of Water on the Slide**
Use your loop or a dropper to place a small amount of water containing the specimens on a clean slide.

PROCEDURE

1. Make wet mount preparations of the living specimens as illustrated in Figures 10-1 and 10-2. You may wish to add a drop of methylene blue to stain the organisms. (If you don't stain them, reduce the light using the iris diaphragm to improve contrast.) A drop of methyl cellulose may also be useful as this slows down the fast swimmers. Observe the following structures on each organism. Sketch and label these in the spaces provided.

    Amoeba: nucleus, pseudopods, vacuoles
    Paramecium: macronucleus, micronucleus, cilia, oral groove, contractile vacuole
    Euglena: nucleus, flagellum

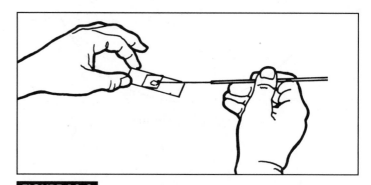

FIGURE 10-2   **Lower the Cover Glass**
Use your loop to gently lower the cover glass onto the water drop. Avoid trapping air bubbles under the cover glass.

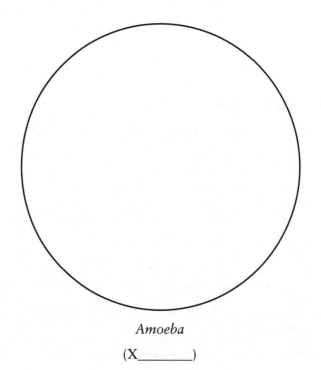

Amoeba

(X_____)

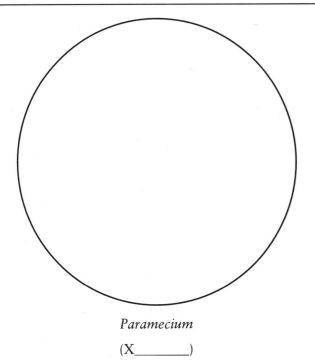

Paramecium

(X_____)

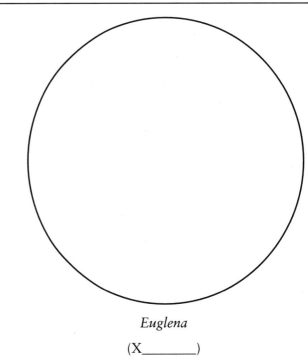

Euglena

(X_____)

2. Obtain prepared slides of the protozoan pathogens and observe them under appropriate magnification. You should observe the assigned structures on each organism. (Many of these slides are made from patient samples, so there will be a lot of other material on the slide besides the desired organism. You must search carefully and with patience.) Sketch and label these in the spaces provided.

Entamoeba histolytica
   trophozoite: pseudopods; nucleus with small, central karyosome and beaded chromatin; ingested erythrocytes
   cysts: multiple nuclei (up to four) with karyosomes and chromatin as in the trophozoite; possibly cytoplasmic chromatoidal bars

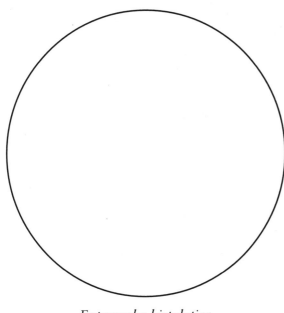

Entamoeba histolytica
trophozoite

(X_____)

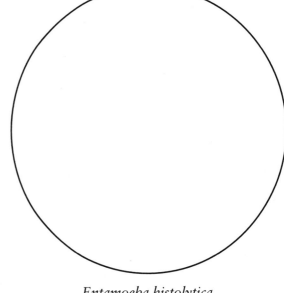

Entamoeba histolytica
cyst

(X_____)

Entamoeba coli

   trophozoite: same as *E. histolytica* except with eccentric karyosome and un-
      clumped chromatin

   cysts: up to 8 nuclei (more than 4 is enough to distinguish it from *E. histolytica*)
      that are the same as in the troph; possibly cytoplasmic chromatoidal bars

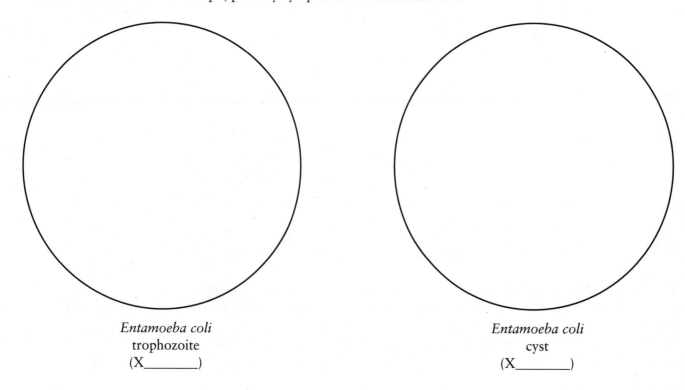

Entamoeba coli
trophozoite
(X_____)

Entamoeba coli
cyst
(X_____)

Balantidium coli

   trophozoite: elongated shape, cilia, macronucleus, possibly micronucleus

   cyst: spherical shape with multiple nuclei

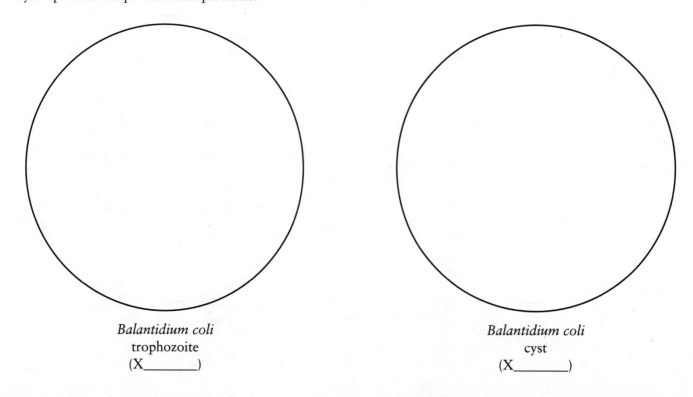

Balantidium coli
trophozoite
(X_____)

Balantidium coli
cyst
(X_____)

Giardia lamblia
   trophozoite: oval shape, flagella (four pairs), nuclei (two), median bodies (two)
   cyst: multiple nuclei (four), median bodies (four)

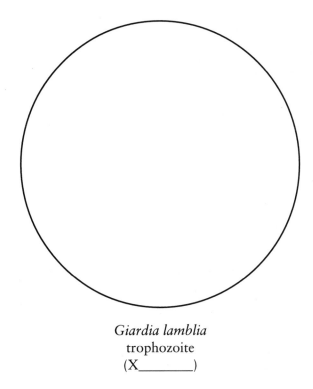

Giardia lamblia
trophozoite
(X_____)

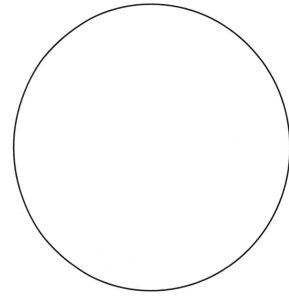

Giardia lamblia
cyst
(X_____)

Trichomonas vaginalis trophozoite: nucleus, flagella (four)
Trypanosoma spp.: nucleus, flagellum, undulating membrane

Trichomonas vaginalis
trophozoite
(X_____)

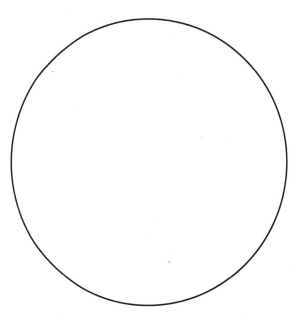

Trypanosoma spp.
(X_____)

Plasmodium falciparum.: ring stage, mature trophozoite, schizont, male gametocyte, female gametocyte

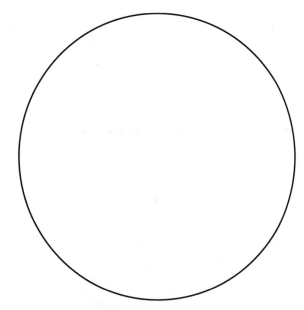

Plasmodium falciparum
ring and mature trophozoites
(X_____)

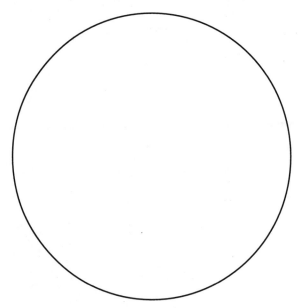

Plasmodium falciparum
schizont
(X_____)

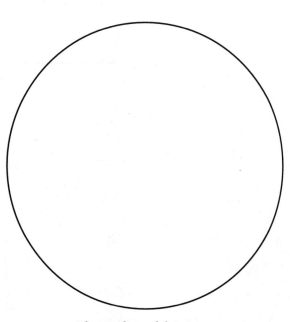

Plasmodium falciparum
gametocytes
(X_____)

REFERENCES

Baron, Ellen Jo, Lance R. Peterson and Sydney M. Finegold. 1994. Chapter 45 in *Bailey and Scott's Diagnostic Microbiology, 9th Ed.* Mosby-Yearbook, St. Louis, MO.

Garcia, Lynne S., Alexander J. Sulzer, George Healy, Katharine K. Grady, and David A. Bruckner. 1995. Chapter 104 in *Manual of Clinical Microbiology, 6th Ed.*, edited by Patrick R. Murray, Ellen Jo Baron, Michael A. Pfaller, Fred C. Tenover, and Robert H. Yolken. American Society for Microbiology, Washington, D.C.

Healy, George R. and Lynne S. Garcia. 1995. Chapter 106 in *Manual of Clinical Microbiology, 6th Ed.*, edited by Patrick R. Murray, Ellen Jo Baron, Michael A. Pfaller, Fred C. Tenover, and Robert H. Yolken. American Society for Microbiology, Washington, D.C.

Koneman, Elmer W., Stephen D. Allen, William M. Janda, Paul C. Schreckenberger and Washington C. Winn, Jr. 1997. Chapter 20 in *Color Atlas and Textbook of Diagnostic Microbiology, 5th Ed.* J.B. Lippincott Company, Philadelphia, PA.

Lee, John J., Seymour H. Hutner, and Eugene C. Bovee. 1985. *Illustrated Guide to the Protozoa.* Society of Protozoologists, Lawrence, KS.

Markell, Edward K., Marietta Voge and David T. John. 1992. *Medical Parasitology, 7th Ed.* W.B. Saunders Company, Philadelphia, PA.

Murray, Patrick R., Ellen Jo Baron, Michael A. Pfaller, Fred C. Tenover, and Robert H. Yolken. 1995. Chapters. 104 and 106 in *Manual of Clinical Microbiology, 6th Ed.* American Society for Microbiology, Washington, D.C.

Exercise 10–3

Helminth Parasites

MATERIALS

Prepared slides of:

Ascaris lumbricoides eggs in a fecal smear
Capillaria hepatica in liver section
Clonorchis sinensis in a fecal smear
Dipylidium caninum eggs in a fecal smear
Echinococcus granulosus hydatid cyst in section
Enterobius vermicularis eggs in a fecal smear
Hookworm *Ancylostoma duodenale* or *Necator americanus* eggs in a fecal smear
Hymenolepis nana eggs in a fecal smear
Onchocerca volvulus in a nodule in section
Paragonimus westermani eggs in a fecal smear
Schistosoma mansoni eggs in a fecal smear
Strongyloides stercoralis rhabditiform larva in a fecal smear
Taenia solium proglottid — W.M.
Taenia solium scolex W.M.
Taenia spp. eggs in a fecal smear
Trichinella spiralis larvae in skeletal muscle — W.M.
Wuchereria bancrofti microfilariae in a blood smear

PROCEDURE

Photographic Atlas Reference Page 164

1. Observe the prepared slides provided of the helminth specimens in fecal smears and other tissues. Scanning on medium power is best for most preparations, then move to high dry or oil immersion to see detail. Most egg specimens are in fecal smears, so there will be a lot of other material on the slide besides the eggs. You must search carefully and with patience. Sketch them in the spaces provided and measure the egg dimensions. Be able to identify each to species if shown an unlabeled specimen.

Trematodes

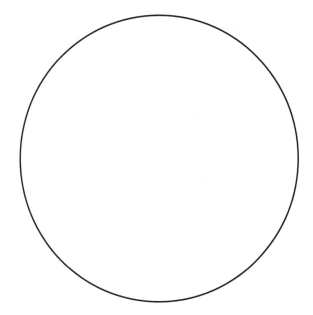

Clonorchis sinensis
egg
(X_____)

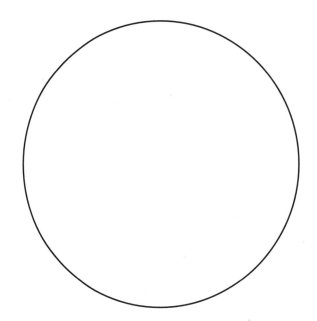

Paragonimus westermani
egg
(X_____)

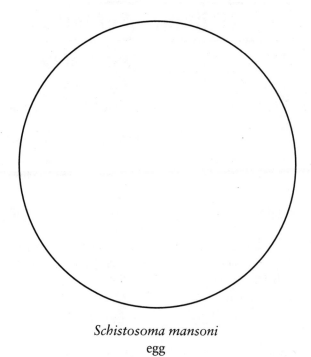

Schistosoma mansoni
egg
(X_____)

Cestodes

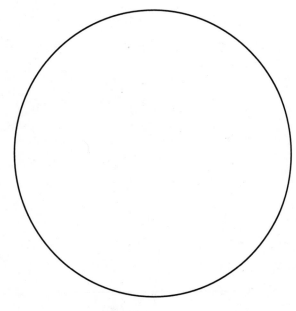

Dipylidium caninum
egg
(X_____)

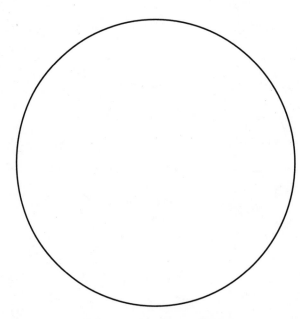

Echinococcus granulosus
protoscolices in a hydatid cyst
(X_____)

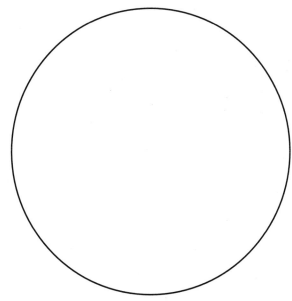

Hymenolepis nana
onchosphere with filaments
(X_____)

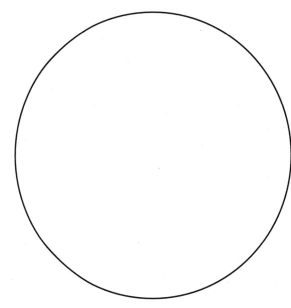

Taenia spp.
egg
(X_____)

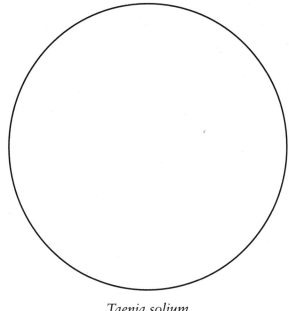

Taenia solium
scolex
(X_____)

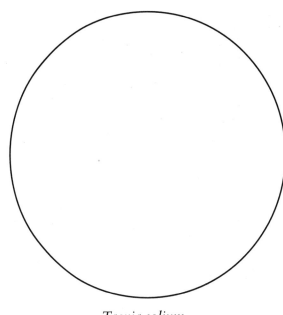

Taenia solium
proglittid
(X_____)

Nematodes

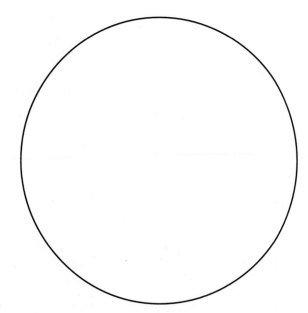

Ascaris lumbricoides
eggs with mammillations
(X_____)

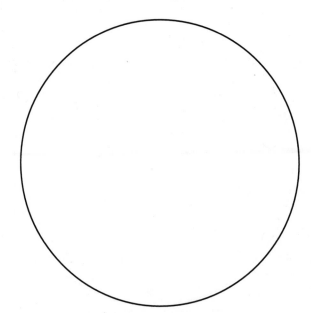

Capillaria hepatica
egg in liver
(X_____)

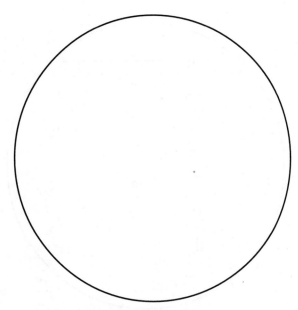

Enterobius vermicularis
egg
(X_____)

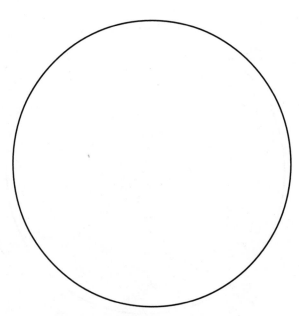

Hookworm *(Ancylostoma* or *Necator)*
egg
(X_____)

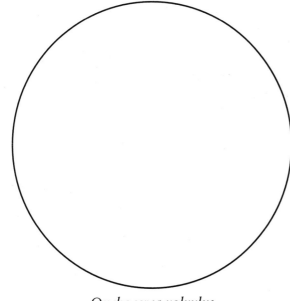

Onchocerca volvulus
larvae in nodule
(X_____)

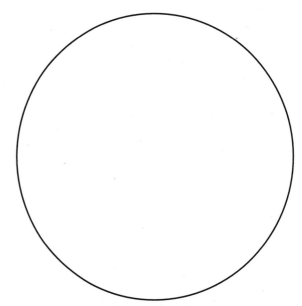

Strongyloides stercoralis
rhabditiform larva
(X_____)

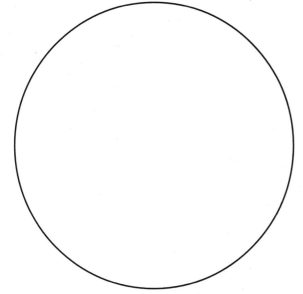

Trichinella spiralis
larvae in skeletal muscle
(X_____)

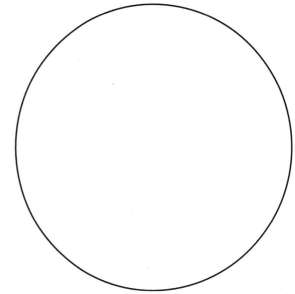

Wuchereria bancrofti
microfilariae in blood
(X_____)

REFERENCES

Ash, Lawrence R., and Thomas C. Orihel. 1995. Chapter 107 in *Manual of Clinical Microbiology, 6th Ed.,* edited by Patrick R. Murray, Ellen Jo Baron, Michael A. Pfaller, Fred C. Tenover, and Robert H. Yolken. American Society for Microbiology, Washington, D.C.

Baron, Ellen Jo, Lance R. Peterson and Sydney M. Finegold. 1994. Chapter 45 in *Bailey and Scott's Diagnostic Microbiology, 9th Ed.* Mosby-Yearbook, St. Louis, MO.

Koneman, Elmer W., Stephen D. Allen, William M. Janda, Paul C. Schreckenberger and Washington C. Winn, Jr. 1997. Chapter 20 in *Color Atlas and Textbook of Diagnostic Microbiology, 5th Ed.* J.B. Lippincott Company, Philadelphia, PA.

Orihel, Thomas C., and Lawrence R. Ash. 1995. Chapter 108 in *Manual of Clinical Microbiology, 6th Ed.,* edited by Patrick R. Murray, Ellen Jo Baron, Michael A. Pfaller, Fred C. Tenover, and Robert H. Yolken. American Society for Microbiology, Washington, D.C.

Roberts, Larry S. and John Janovy, Jr. *Foundations of Parasitology, 5th Ed.* Wm. C. Brown Publishers, Dubuque, IA.

Recipes for Media Used in this Manual

1. Bile Esculin Agar

Beef extract	3.0 g
Peptone	5.0 g
Oxgall	40.0 g
Esculin	1.0 g
Ferric citrate	0.5 g
Agar	15.0 g
Distilled or deionized water	1.0 L

final pH = 6.6 ± 0.2 at 25°C

1. Suspend the ingredients in one liter of distilled or deionized water, mix well and boil to dissolve completely.
2. Dispense 7.0 mL volumes into test tubes and cap loosely.
3. Sterilize in the autoclave at 15 lbs. pressure (121°C) for 15 minutes.
4. Remove from the autoclave, slant and allow to cool to room temperature.

2. Blood Agar

Infusion from beef heart (solids)	2.0 g
Pancreatic digest of casein	13.0 g
Sodium chloride	5.0 g
Yeast extract	5.0 g
Agar	15.0 g
Defibrinated sheep blood	50.0 mL
Distilled or deionized water	1.0 L

final pH = 7.3 ± 0.2 at 25°C

1. Suspend, mix and boil the dry ingredients in one liter distilled or deionized water. This is blood agar base.
2. Cover loosely and sterilize in the autoclave at 15 lbs. pressure (121°C) for 15 minutes.
3. Remove from the autoclave and cool to 45°C.
4. Aseptically add the sterile, room temperature sheep blood to the blood agar base and mix well.
5. Pour into sterile Petri dishes and allow to cool to room temperature.

3. Brilliant Green Lactose Bile Broth

Peptone	10.0 g
Lactose	10.0 g
Oxgall	20.0 g
Brilliant green dye	0.0133 g
Distilled or deionized water	1.0 L

final pH = 7.2 ± 0.2 at 25°C

1. Suspend, mix and heat the ingredients in one liter of distilled or deionized water until completely dissolved.
2. Dispense 10.0 mL portions into test tubes,
3. Place an inverted Durham tube in each broth and cap loosely.
4. Sterilize in the autoclave at 15 lbs. pressure (121°C) for 15 minutes.
5. Remove the media from the autoclave and allow it to cool before inoculating.

4. Citrate Agar (Simmons)

Ammonium dihydrogen phosphate	1.0 g
Dipotassium phosphate	1.0 g
Sodium chloride	5.0 g
Sodium citrate	2.0 g
Magnesium sulfate	0.2 g
Agar	15.0 g
Bromthymol blue	0.08 g
Distilled or deionized water	1.0 L

final pH = 6.9 ± 0.2 at 25°C

1. Suspend the ingredients in one liter distilled or deionized water, mix well and boil to dissolve completely.
2. Dispense 7.0 mL portions into test tubes and cap loosely.
3. Sterilize in the autoclave at 15 lbs pressure (121°C) for 15 minutes.
4. Remove from the autoclave, slant and allow to cool to room temperature.

5. Complete Medium (Ames Test)

Beef extract	3.0 g
Peptone	5.0 g
Sodium chloride	5.0 g
Agar	20.0 g
Distilled or deionized water	1.0 L

1. Suspend, mix and boil the ingredients in one liter of distilled or deionized water until completely dissolved.

2. Cover loosely and sterilize in the autoclave at 15 lbs. pressure (121°C) for 15 minutes.

3. Remove from the autoclave and cool slightly.

4. Aseptically pour into sterile petri dishes (15 mL/plate) and allow to cool to room temperature.

6. Decarboxylase Medium (Møller)

Peptone	5.0 g
Beef extract	5.0 g
Glucose (dextrose)	0.5 g
Bromcresol purple	0.01 g
Cresol red	0.005 g
Pyridoxal	0.005 g
L-Lysine, L-Ornithine, or L-Arginine	10.0 g
Distilled or deionized water	1.0 L

final pH = 6.0 ± 0.2 at 25°C loosely.

1. Suspend and heat the ingredients in one liter of distilled or deionized water until completely dissolved. (Use only one of the listed L-amino acids.)

2. Adjust pH by adding NaOH if necessary.

3. Dispense 7.0 mL volumes into test tubes and cap

4. Sterilize in the autoclave at 15 lbs. pressure (121°C) for 10 minutes.

5. Remove from the autoclave and allow to cool to room temperature.

7. Desoxycholate Agar (Modified Leifson)

Peptone	10.0 g
Lactose	10.0 g
Sodium desoxycholate	1.0 g
Sodium chloride	5.0 g
Dipotassium phosphate	2.0 g
Ferric citrate	1.0 g
Sodium citrate	1.0 g
Agar	16.0 g
Neutral red	0.033 g
Distilled or deionized water	1.0 L

final pH = 7.3 ± 0.2 at 25°C

1. Heat and stir the ingredients in one liter of distilled or deionized water. Boil for 1 minute to make certain they are completely dissolved.

2. When cooled to 50°C, pour into sterile plates.

3. Allow to cool to room temperature.

8. DNase Test Agar with Methyl Green

Tryptose	20.0 g
Deoxyribonucleic acid	2.0 g
Sodium chloride	5.0 g
Agar	15.0 g
Methyl green	0.05 g
Distilled or deionized water	1.0 L

final pH = 7.3 ± 0.2 at 25°C

1. Suspend, mix and boil the ingredients in one liter of distilled or deionized water until completely dissolved.

2. Cover loosely and sterilize in the autoclave at 15 lbs. pressure (121°C) for 15 minutes.

3. Aseptically pour into sterile Petri dishes (15 mL/plate) and allow to cool to room temperature.

9. EC Broth

Tryptose	20.0 g
Lactose	5.0 g
Dipotassium phosphate	4.0 g
Monopotassium phosphate	1.5 g
Sodium chloride	5.0 g
Distilled or deionized water	1.0 L

final pH = 6.9 ± 0.2 at 25°C.

1. Suspend, mix and heat the ingredients in one liter of distilled or deionized water until completely dissolved.

2. Dispense 10.0 mL portions into test tubes,

3. Place an inverted Durham tube in each broth and cap loosely.

4. Sterilize in the autoclave at 15 lbs. pressure (121°C) for 15 minutes.

5. Remove the media from the autoclave and allow it to cool before inoculating.

10. Eosin Methylene Blue Agar (Levine)

Peptone	10.0 g
Lactose	10.0 g*
Dipotassium phosphate	2.0 g
Agar	15.0 g
Eosin Y	0.4 g
Methylene blue	0.065 g
Distilled or deionized water	1.0 L

final pH = 7.1 ± 0.2 at 25°C

1. Mix and heat the ingredients in one liter of distilled or deionized water until they are completely dissolved.

2. Autoclave for 15 minutes at 15 lbs. pressure (121°C).

3. When cooled to 50°C, pour into sterile plates.

4. Allow to cool to room temperature.

---

* An alternate recipe replaces the 10.0 g of lactose with 5.0 g of lactose and 5.0 g of sucrose.

11. Hektoen Agar

Yeast extract	3.0 g
Peptic digest of animal tissue	12.0 g
Lactose	12.0 g
Sucrose	12.0 g
Salicin	2.0 g
Bile salts	9.0 g
Sodium chloride	5.0 g
Sodium thiosulfate	5.0 g
Ferric ammonium citrate	1.5 g
Bromthymol blue	0.064 g
Acid fuchsin	0.1 g
Agar	13.5 g
Distilled or deionized water	1.0 L

final pH = 7.6 ± 0.2 at 25°C

1. Mix and heat the ingredients in one liter of distilled or deionized water until they are dissolved. Boil for 1 minute to make certain they are completely dissolved.

2. Do not autoclave.

3. When cooled to 50°C, pour into sterile plates.

4. Allow to cool to room temperature with lids slightly open.

12. Kligler's Iron Agar

Beef extract	3.0 g
Yeast extract	3.0 g
Peptone	15.0 g
Proteose peptone	5.0 g
Lactose	10.0 g
Dextrose (glucose)	1.0 g
Ferrous sulfate	0.2 g
Sodium chloride	5.0 g
Sodium thiosulfate	0.3 g
Agar	12.0 g
Phenol red	0.024 g
Distilled or deionized water	1.0 L

final pH = 7.4 ± 0.2 at 25°C

1. Suspend, mix and boil the ingredients in one liter of distilled or deionized water until completely dissolved.

2. Transfer 7.0 mL portions to test tubes and cap loosely.

3. Sterilize in the autoclave at 15 lbs. pressure (121°C) for 15 minutes.

4. Remove from the autoclave and slant in such a way as to form a deep butt.

5. Allow to cool to room temperature.

13. Lauryl Tryptose Broth

Tryptose	20.0 g
Lactose	5.0 g
Dipotassium phosphate	2.75 g
Monopotassium phosphate	2.75 g
Sodium chloride	5.0 g
Sodium lauryl sulfate	0.1 g
Distilled or deionized water	1.0 L

final pH = 6.8 ± 0.2 at 25°C

1. Suspend, mix and heat the ingredients in one liter of distilled or deionized water until completely dissolved.
2. Dispense 10.0 mL portions into test tubes,
3. Place an inverted Durham tube in each broth and cap loosely.
4. Sterilize in the autoclave at 15 lbs. pressure (121°C) for 15 minutes.
5. Remove the media from the autoclave and allow it to cool before inoculating.

14. Litmus Milk Medium

Skim milk	100.0 g
Azolitmin	0.5 g
Sodium sulfite	0.5 g
Distilled or deionized water	1.0 L

final pH = 6.5 ± 0.2 at 25°C

1. Suspend and mix the ingredients in one liter of deionized or distilled water and heat to approximately 50°C to dissolve completely.
2. Transfer 7.0 mL portions to test tubes and cap loosely.
3. Sterilize in the autoclave at 113–115°C for 20 minutes.
4. Remove from the autoclave and allow to cool to room temperature.

15. Lysine Iron Agar

Peptone	5.0 g
Yeast Extract	3.0 g
Dextrose	1.0 g
L-Lysine hydrochloride	10.0 g
Ferric ammonium citrate	0.5 g
Sodium thiosulfate	0.04 g
Bromcresol purple	0.02 g
Agar	15.0 g

final pH = 6.7 ± 0.2 at 25°C

1. Suspend and mix the ingredients in one liter of deionized or distilled water and boil to completely dissolve.
2. Transfer 8.0 mL portions to test tubes and cap loosely.
3. Sterilize in the autoclave at 121°C for 15 minutes.
4. Remove from the autoclave and slant in such a way as to produce a deep butt. Allow the media to cool to room temperature.

16. MacConkey Agar

Pancreatic digest of gelatin	17.0 g
Pancreatic digest of casein	1.5 g
Peptic digest of animal tissue	1.5 g
Lactose	10.0 g
Bile salts	1.5 g
Sodium chloride	5.0 g
Neutral red	0.03 g
Crystal violet	0.001
Agar	13.5 g
Distilled or deionized water	1.0 L

final pH = 7.1 ± 0.2 at 25°C

1. Mix and heat the ingredients in one liter of distilled or deionized water until they are dissolved. Boil for 1 minute to make certain they are completely dissolved.
2. Autoclave for 15 minutes at 15 lbs. pressure (121°C).
3. When cooled to 50°C, pour into sterile plates.
4. Allow to cool to room temperature.

17. Mannitol Salt Agar

Beef extract	1.0 g
Peptone	10.0 g
Sodium chloride	75.0 g
D-Mannitol	10.0 g
Phenol red	0.025 g
Agar	15.0 g
Distilled or deionized water	1.0 L

final pH = 7.4 ± 0.2 at 25°C

1. Suspend the ingredients in one liter of distilled or deionized water and mix. Boil one minute to completely dissolve ingredients.

2. Autoclave for 15 minutes at 15 lbs. pressure (121°C).

3. When cooled to 50°C, pour into sterile plates.

4. Allow plates to cool to room temperature.

18. Milk Agar

Beef extract	3.0 g
Peptone	5.0 g
Agar	15.0 g
Powdered nonfat milk	100.0 g
Distilled or deionized water	1.0 L

final pH = 7.2 ± 0.2 at 25°C

1. Suspend the powdered milk in 500.0 mL of distilled or deionized water in a one liter flask, mix well and cover loosely.

2. Suspend the remainder of the ingredients in 500.0 mL of deionized water in a one liter flask, mix well, boil to dissolve completely and cover loosely.

3. Sterilize in the autoclave at 113–115°C for 20 minutes.

4. Remove from the autoclave, allow to cool slightly, then aseptically pour the milk solution into the agar solution and mix *gently* (to prevent foaming).

5. Aseptically pour into sterile Petri dishes (15 mL/plate).

6. Allow to cool to room temperature.

19. Minimal Medium (Ames Test)

Dextrose (glucose)	20.0 g
50x Vogel-Bonner salts	20.0 mL
Histidine	0.00016 g
Biotin	0.00025 g
Agar	20.0 g
Distilled or deionized water	1.0 L

1. Add 1.6 mg histidine to 10.0 mL distilled or deionized water and filter sterilize.

2. Add 2.5 mg biotin to 10.0 mL distilled or deionized water and filter sterilize.

3. Prepare the 50x Vogel-Bonner salts solution by adding the ingredients to *just enough* water to dissolve them while heating and stirring. After the ingredients are dissolved add enough water to bring the total volume up to exactly one liter.

4. Suspend, mix and boil the agar in 500.0 mL of distilled or deionized water until completely dissolved.

5. Suspend and mix the dextrose in 500.0 mL of distilled or deionized water until completely dissolved.

6. Cover the agar and dextrose containers loosely and sterilize in the autoclave at 121°C for 15 minutes.

7. Remove from the autoclave and allow to cool to 80°C.

8. Aseptically add 1.0 mL histidine solution, 1.0 mL biotin solution, and 20 mL 50x Vogel-Bonner salts to the glucose solution and mix well.

9. Add the glucose solution to the agar solution, mix well and aseptically pour into sterile Petri dishes (15 mL/plate).

20. Motility Test Medium

Beef extract	3.0 g
Pancreatic digest of gelatin	10.0 g
Sodium chloride	5.0 g
Agar	4.0 g
Triphenyltetrazolium chloride (TTC)	0.05 g
Distilled or deionized water	1.0 L

final pH = 7.3 ± 0.2 at 25°C

1. Suspend the ingredients in one liter of distilled or deionized water, mix well and boil to dissolve completely.
2. Dispense 7.0 mL portions into test tubes and cap loosely.
3. Sterilize in the autoclave at 15 lbs. pressure (121°C) for 15 minutes.
4. Remove from the autoclave and allow to cool in the upright position.

21. MRVP Broth

Buffered peptone	7.0 g
Dipotassium phosphate	5.0 g
Dextrose (glucose)	5.0 g
Distilled or deionized water	1.0 L

final pH = 6.9 ± 0.2 at 25°C

1. Suspend the ingredients in one liter of deionized or distilled water, mix well and warm until completely dissolved.
2. Transfer 7.0 mL portions to test tubes and cap loosely.
3. Sterilize in the autoclave at 15 lbs. pressure (121°C) for 15 minutes.
4. Remove from the autoclave and allow to cool to room temperature.

22. Mueller-Hinton II Agar

Beef extract	2.0 g
Acid hydrolysate of casein	17.5 g
Starch	1.5 g
Agar	17.0 g
Distilled or deionized water	1.0 L

final pH = 7.3 ± 0.1 at 25°C

1. Suspend the ingredients in one liter of distilled or deionized water, mix well and boil to dissolve completely.
2. Cover loosely and sterilize in the autoclave at 121°C (15 lbs.) for 15 minutes.
3. Remove from the autoclave, allow to cool slightly.
4. Aseptically pour into sterile Petri dishes to a depth of 4 mm.
5. Allow to cool to room temperature.

23. Nutrient Agar

Beef extract	3.0 g
Peptone	5.0 g
Agar	15.0 g
Distilled or deionized water	1.0 L

final pH = 6.8 ± 0.2 at 25°C

Plates

1. Suspend the ingredients in one liter of distilled or deionized water, mix well and boil to dissolve completely.
2. Cover loosely and sterilize in the autoclave at 15 lbs. pressure (121°C) for 15 minutes.
3. Remove from the autoclave, allow to cool slightly and aseptically pour into sterile Petri dishes (15 mL/plate).
4. Allow to cool to room temperature.

Tubes

1. Suspend the ingredients in one liter of distilled or deionized water, mix well, and boil until fully dissolved.
2. Dispense 10 mL portions into test tubes (7 mL for slants) and cap loosely.
3. Autoclave for 15 minutes at 121°C to sterilize the medium.
4. Cool to room temperature with the tubes in an upright position for agar deep tubes. Cool with the tubes on an angle for agar slants.

24. Nutrient Broth

Beef extract	3.0 g
Peptone	5.0 g
Distilled or deionized water	1.0 L

final pH = 6.8 ± 0.2 at 25°C

1. Suspend the ingredients in one liter of distilled or deionized water. Agitate and heat slightly (if necessary) to dissolve completely.

2. Dispense 7.0 mL portions into test tubes and cap loosely.

3. Autoclave for 15 minutes at 121°C to sterilize the medium.

25. Nutrient Gelatin

Beef Extract	3.0 g
Peptone	5.0 g
Gelatin	120.0 g
Distilled or deionized water	1.0 L

final pH = 6.8 ± 0.2 at 25°C

1. Slowly add the ingredients to one liter of distilled or deionized water while stirring.

2. Warm to 50°C and maintain temperature until completely dissolved.

3. Dispense 7.0 mL volumes into test tubes and cap loosely.

4. Sterilize in the autoclave at 15 lbs. pressure (121°C) for 15 minutes.

5. Remove from the autoclave immediately and allow to cool to room temperature in the upright position.

26. OF Basal Medium

Pancreatic digest of casein	2.0 g
Sodium chloride	5.0 g
Dipotassium phosphate	0.3 g
Agar	2.5 g
Bromthymol blue	0.03 g
Distilled or deionized water	1.0 L

final pH = 6.8 ± 0.1 at 25°C

OF Carbohydrate Solution

Carbohydrate (glucose, lactose, sucrose)	1.0 g
Distilled or deionized water	<10.0 mL*

1. Suspend the ingredients, *without the carbohydrate*, in one liter of distilled or deionized water, mix well and boil to dissolve completely. This is basal medium.

2. Divide the medium into ten aliquots of 100.0 mL each.

3. Cover loosely and sterilize in the autoclave at 121°C for 15 minutes.

4. Cool to 50°C.

5. Prepare 10% solutions of carbohydrates to be tested (below).

6. Sterilize in the autoclave at 118°C for 10 minutes.

7. Aseptically add 10.0 mL sterile carbohydrate solution to a 50°C basal medium aliquot and mix well.

8. Aseptically transfer 7.0 mL volumes to sterile test tubes and allow to cool.

---

* Total solution volume is 10.0 mL

27. Phenol Red (Carbohydrate) Broth

Pancreatic digest of casein	10.0 g
Sodium chloride	5.0 g
Carbohydrate (glucose, lactose, sucrose)	5.0 g
Phenol red	0.018 g
Distilled or deionized water	1.0 L

final pH = 7.3 ± 0.2 at 25°C

1. Suspend the ingredients in one liter of distilled or deionized water, mix well and warm slightly to dissolve completely.

2. Dispense 7.0 mL volumes into test tubes.

3. Insert inverted Durham tubes into the test tubes and cap loosely.

4. Sterilize in the autoclave at 116–118°C for 15 minutes.

5. Remove from the autoclave and allow to cool to room temperature.

28. Phenylalanine Deaminase Agar

DL-Phenylalanine	2.0 g
Yeast extract	3.0 g
Sodium chloride	5.0 g
Sodium phosphate	1.0 g
Agar	12.0 g
Distilled or deionized water	1.0 L

final pH = 7.3 ± 0.2 at 25°C

1. Suspend the ingredients in one liter of distilled or deionized water, mix well and boil to dissolve completely.

2. Dispense 7.0 mL volumes into test tubes and cap loosely.

3. Sterilize in the autoclave at 15 lbs pressure (121°C) for 10 minutes.

4. Remove from the autoclave, slant and allow to cool to room temperature.

29. Phenylethyl Alcohol Agar

Tryptose	10.0 g
Beef extract	3.0 g
Sodium chloride	5.0 g
Phenylethyl alcohol	2.5 g
Agar	15.0 g
Distilled or deionized water	1.0 L

final pH = 7.3 ± 0.2 at 25°C

1. Suspend the ingredients in one liter of distilled or deionized water. Boil one minute to completely dissolve ingredients.

2. Autoclave for 15 minutes at 15 lbs. pressure (121°C).

3. When cooled to 50°C, pour into sterile plates.

4. Allow plates to cool to room temperature.

30. Purple Broth

Peptone	10.0 g
Beef extract	1.0 g
Sodium chloride	5.0 g
Bromcresol purple	0.02 g
Carbohydrate (glucose, lactose or sucrose)	10.0 g
Distilled or deionized water	1.0 L

final pH = 6.8 ± 0.2 at 25°C

1. Suspend the ingredients in one liter of distilled or deionized water, mix well and warm slightly to dissolve completely.

2. Dispense 9.0 mL volumes into test tubes.

3. Insert inverted Durham tubes into the test tubes and cap loosely.

4. Sterilize in the autoclave at 118°C for 15 minutes.

5. Remove from the autoclave and allow to cool to room temperature.

31. Saline Agar — Double Gel Immunodiffusion

Sodium chloride	10.0 g
Agar	20.0 g
Distilled or deionized water	1.0 L

1. Suspend the ingredients in one liter distilled or deionized water, mix well and boil to dissolve completely.

2. Pour into Petri dishes to a depth of 3 mm. Do not replace the lids until the agar has solidified and cooled to room temperature.

32. SIM (Sulfur-Indole-Motility) Medium

Pancreatic digest of casein	20.0 g
Peptic digest of animal tissue	6.1 g
Ferrous ammonium sulfate	0.2 g
Sodium thiosulfate	0.2 g
Agar	3.5 g
Distilled or deionized water	1.0 L

final pH = 7.3 ± 0.2 at 25°C

1. Suspend the ingredients in one liter of distilled or deionized water, mix well and boil to dissolve completely.

2. Dispense 7.0 mL volumes into test tubes and cap loosely.

3. Sterilize in the autoclave at 15 lbs pressure (121°C) for 15 minutes.

4. Remove from the autoclave and allow to cool to room temperature

33. Snyder Test Medium

Pancreatic digest of casein	13.5 g
Yeast extract	6.5 g
Dextrose	20.0 g
Sodium chloride	5.0 g
Agar	16.0 g
Bromcresol green	0.02 g
Distilled or deionized water	1.0 L

final pH = 4.8 ± 0.2 at 25°C

1. Suspend the ingredients in one liter distilled or deionized water, mix well and boil to dissolve completely.

2. Transfer 7.0 mL portions to test tubes and cap loosely.

3. Sterilize in the autoclave at 118–121°C for 15 minutes.

4. Remove from the autoclave and place in a hot water bath set at 45–50°C. Allow at least 30 minutes for the agar temperature to equilibrate before beginning the exercise.

34. Soft Agar

Beef extract	3.0 g
Peptone	5.0 g
Sodium chloride	5.0 g
Tryptone	2.5 g
Yeast extract	2.5 g
Agar	7.0 g
Distilled or deionized water	1.0 L

1. Suspend, mix and boil the ingredients in one liter of distilled or deionized water until completely dissolved.

2. Transfer 2.5 mL portions to test tubes and cap loosely.

3. Sterilize in the autoclave at 15 lbs. pressure (121°C) for 15 minutes.

4. Remove from the autoclave and place in a hot water bath set at 45°C. Allow 30 minutes for the agar temperature to equilibrate.

35. Spirit Blue Agar

Tryptone	10.0 g
Yeast extract	5.0 g
Agar	20.0 g
Spirit blue	0.15 g
Olive oil	100.0 mL
Distilled or deionized water	1.0 L

final pH = 6.8 ± 0.2 at 25°C

1. Suspend, mix and boil all the ingredients except the olive oil in one liter of distilled or deionized water until completely dissolved.

2. Cover loosely and sterilize in the autoclave at 121°C for 15 minutes. Sterilize the olive oil separately at the same time.

3. Remove from the autoclave and allow to cool to approximately 50°C.

4. Blend the ingredients together in a sterile blender for one minute on "High."

5. Poor into sterile Petri dishes and allow to cool.

6. Wait 24 hours before inoculating.

36. Starch Agar

Beef extract	3.0 g
Soluble starch	10.0 g
Agar	12.0 g
Distilled or deionized water	1.0 L

final pH = 7.5 ± 0.2 at 25°C

1. Suspend the ingredients in one liter of distilled or deionized water, mix well and boil to dissolve completely.

2. Sterilize in the autoclave at 15 lbs pressure (121°C) for 15 minutes.

3. Remove from the autoclave and allow to cool slightly.

4. Aseptically pour into sterile Petri dishes (15 mL per plate).

37. Thioglycolate Broth

Yeast Extract	5.0 g
Casitone	15.0 g
Dextrose (glucose)	5.5 g
Sodium chloride	2.5 g
Sodium thioglycolate	0.5 g
L-Cystine	0.5 g
Agar	0.75 g
Resazurin	0.001 g
Distilled or deionized water	1.0 L

final pH = 7.1 ± 0.2 at 25°C

1. Suspend the ingredients in one liter of distilled or deionized water. Boil to completely dissolve them.

2. Dispense 10.0 mL into sterile test tubes.

3. Autoclave for 15 minutes at 15 lbs. pressure (121°C) to sterilize.

38. Tributyrin Agar

Beef extract	1.5 g
Peptone	2.5 g
Agar	7.5 g
Tributyrin oil	5.0 mL
Distilled or deionized water	500.0 mL

final pH = 6.0 ± 0.2 at 25°C

1. Suspend the dry ingredients in 500.0 mL of deionized water, mix well and boil to dissolve completely.

2. Cover loosely and sterilize together with the tube of tributyrin oil in the autoclave at 15 lbs pressure (121°C) for 15 minutes.

3. Remove from the autoclave and aseptically pour agar mixture into a sterile glass blender.

4. Aseptically add the tributyrin oil to the agar mixture and blend on "High" for 1 minute.

5. Aseptically pour into sterile Petri dishes (15 mL/plate).

39. Triple Sugar Iron Agar

Beef extract	3.0 g
Yeast extract	3.0 g
Peptone	15.0 g
Proteose peptone	5.0 g
Dextrose (glucose)	1.0 g
Lactose	10.0 g
Sucrose	10.0 g
Ferrous sulfate	0.2 g
Sodium chloride	5.0 g
Sodium thiosulfate	0.3 g
Agar	12.0 g
Phenol red	0.024 g
Distilled or deionized water	1.0 L

final pH = 7.4 ± 0.2 at 25°C

1. Suspend the ingredients in one liter of distilled or deionized water, mix well and boil to dissolve completely.
2. Transfer 7.0 mL portions to test tubes and cap loosely.
3. Sterilize in the autoclave at 121°C for 15 minutes.
4. Slant in such a way as to form a deep butt.
5. Allow to cool to room temperature.

40. Tryptic Nitrate Medium

Tryptose	20.0 g
Dextrose	1.0 g
Disodium phosphate	2.0 g
Potassium nitrate	1.0 g
Agar	1.0 g
Distilled or deionized water	1.0 L

final pH = 7.2 ± 0.2 at 25°C

1. Suspend the ingredients in one liter of deionized or distilled water, mix well and boil until completely dissolved.
2. Transfer 10.0 mL portions to test tubes and cap loosely.
3. Sterilize in the autoclave at 15 lbs. pressure (121°C) for 15 minutes.
4. Remove from the autoclave and allow to cool to room temperature.

41. Urease Agar

Peptone	1.0 g
Dextrose (glucose)	1.0 g
Sodium chloride	5.0 g
Potassium phosphate, monobasic	2.0 g
Agar	15.0 g
Phenol red	0.012 g
Distilled or deionized water	1.0 L

final pH = 6.8 ± 0.2 at 25°C

1. Suspend the agar in 900 mL distilled or deionized water, mix well and boil to dissolve completely.
2. Cover loosely and sterilize by autoclaving at 15 lbs. pressure (121°C) for 15 minutes.
3. Remove from the autoclave and allow to cool to 55°C.
4. Suspend the remaining ingredients in 100 mL distilled or deionized water, mix well and filter sterilize. *Do not autoclave.* This is urease agar base.
5. Aseptically add the urease agar base to the agar solution and mix well.
6. Aseptically transfer 7.0 mL portions to sterile test tubes and cap loosely.
7. Slant in such a way that the agar butt is approximately twice as long as the slant.
8. Allow to cool to room temperature.

42. Urease Broth

Yeast extract	0.1 g
Potassium phosphate, monobasic	9.1 g
Potassium phosphate, dibasic	9.5 g
Urea	20.0 g
Phenol red	0.01 g
Distilled or deionized water	1.0 L

final pH = 6.8 ± 0.2 at 25°C

1. Suspend the ingredients in one liter distilled or deionized water and mix well.
2. Filter sterilize the solution. *Do not autoclave.*
3. Aseptically transfer 1.0 mL volumes to small sterile test tubes and cap loosely.

43. Xylose Lysine Desoxycholate Agar

Xylose	3.5 g
L-Lysine	5.0 g
Lactose	7.5 g
Sucrose	7.5 g
Sodium chloride	5.0 g
Yeast extract	3.0 g
Phenol red	0.08 g
Sodium desoxycholate	2.5 g
Sodium thiosulfate	6.8 g
Ferric ammonium citrate	0.8 g
Agar	13.5 g
Distilled or deionized water	1.0 L

final pH = 7.5 ± 0.2 at 25°C

1. Suspend the ingredients in one liter of distilled or deionized water and mix. Heat only until the medium boils.

2. Cool in a water bath at 50°C.

3. When cooled, pour into sterile plates.

4. Allow to plates to cool to room temperature.

References

Baron, Ellen Jo, Lance R. Peterson and Sydney M. Finegold. 1994. Page 329 in *Bailey & Scott's Diagnostic Microbiology, 9th Ed.* Mosby–Year Book, Inc. St. Louis, Missouri.

DIFCO Laboratories. 1984. *DIFCO Manual, 10th Ed.* DIFCO Laboratories, Detroit, MI.

Koneman, Elmer W., *et al.* 1997. Page 551 and 1299 in *Color Atlas and Textbook of Diagnostic Microbiology, 5th Ed.* Lippincott-Raven Publishers, Philadelphia, PA.

MacFaddin, Jean F. 1980. Page 183 in *Biochemical Tests for Identification of Medical Bacteria, 2nd Ed.* Williams & Wilkins, Baltimore, MD.

Power, David A. and Peggy J. McCuen. 1988. *Manual of BBL® Products and Laboratory Procedures, 6th Ed.* Becton Dickinson Microbiology Systems, Cockeysville, MD.

Recipes for Reagents and Stains Used in this Manual

A. REAGENTS

1. Dilution Tubes — Viable Count, Plaque Assay

NaCl	0.9 g
Distilled or deionized water	$\cong 100.0$ mL

1. Dissolve the NaCl in approximately 90 mL of water.

2. Add water to bring the total volume up to 100.0 mL.

3. Cover loosely and sterilized in the autoclave at 121°C for 15 minutes.

4. Allow to cool to room temperature.

5. Aseptically dispense into sterile test tubes in the amounts of 9.0 mL and 9.9 mL.

2. Direct Count — Staining/Diluting Agents

Agent A

100% saturated crystal violet-ethanol solution	40.0 mL
NaCl	0.9 g
Distilled or deionized water	$\cong 60.0$ mL

1. Dissolve the crystal violet in ethanol and filter.

2. Mix 0.9 g NaCl in approximately 55 mL of distilled or deionized water.

3. Add the NaCl solution to 40 mL of the crystal violet-ethanol solution.

4. Add water to bring the total volume up to 100 mL.

Agent B

Ethanol	40.0 mL
NaCl	0.9 g
Distilled or deionized water	$\cong 60.0$ mL

1. Dissolve 0.9 g NaCl in approximately 55 mL of distilled or deionized water.

2. Add the mixture to 40 mL of ethanol.

3. Add water to bring the total volume up to 100 mL.

3. MR/VP (Methyl Red — Voges-Proskauer Test Reagents)

Methyl Red

Methyl red dye	0.1 g
Ethanol	300.0 mL
Distilled water	$\cong 200.0$ mL

1. Dissolve the dye in the ethanol.

2. Add water to bring the total volume up to 500 mL.

VP Reagent A

α-naphthol	5.0 g
Absolute Ethanol	$\cong 100.0$ mL

1. Dissolve the α-naphthol in approximately 95 mL of water.

2. Add water to bring the total volume up to 100 mL.

VP Reagent B

Potassium hydroxide	40.0 g
Creatine	2.5 g
Distilled water	$\cong 50$ mL

1. Dissolve the potassium hydroxide in approximately 50 mL of water. (**Caution:** This solution is highly concentrated and will become hot as the KOH dissolves. It should be prepared in appropriate glassware on a stirring hot plate.) Allow it to cool to room temperature.

2. Add the creatine.

3. Add water to bring the volume up to 100 mL.

4. McFarland Turbidity Standard (0.5)

Barium chloride ($BaCl_2 \cdot 2\ H_2O$)	1.175 g
Sulfuric acid, concentrated (H_2SO_4)	1.0 mL
Distilled or deionized water	$\cong$ 200.0 mL

Use only clean glassware for this preparation.

1. Pour approximately 90 mL of water into a small erlenmeyer flask.

2. Add the $BaCl_2$ and mix well.

3. Remeasure and add water to bring the total volume up to 100 mL.

4. Again using only clean glassware, add the H_2SO_4 to approximately 90 mL of water.

5. Remeasure and add water to bring the total volume up to 100 mL.

6. Add 0.5 mL of the $BaCl_2$ solution to 99.5 mL of H_2SO_4 and mix well.

7. While keeping the solution well mixed (the barium sulfate will precipitate and settle out) distribute 7 to 10 mL volumes into screw cap test tubes.

5. Methylene Blue Reductase Reagent

Methylene blue dye	8.8 mg
Distilled or deionized water	200.0 mL

6. Nitrate Test Reagents

Reagent A

Sulfanilic acid	1.0 g
5N Acetic acid	125.0 mL

Reagent B

Dimethyl-α-naphthylamine	1.0 g
5N Acetic acid	200.0

7. Oxidase Test Reagent

Tetramethyl-p-phenylenediamine dihydrochloride	1.0 g
Deionized Water	100.0 mL

8. Phenylalanine Deaminase Test Reagent

Ferric chloride	10.0 g
Deionized water	$\cong$ 90.0 mL

1. Dissolve the ferric chloride in approximately 90 mL of distilled or deionized water.

2. Add water to bring the total volume up to 100 mL.

B. STAINS

1. Acid Fast, Cold Stain Reagents (Modified Kinyoun)

Carbolfuchsin

Basic fuchsin	1.5g
Phenol	4.5 g
Ethanol (95%)	5.0 mL
Isopropanol	20.0 mL
Distilled or deionized water	75.0 mL

1. Dissolve the basic fuchsin in the ethanol and add the isopropanol.

2. Mix the phenol in the water.

3. Mix the solutions together and let stand for several days.

4. Filter before use.

Decolorizer

H_2SO_4	1.0 mL
Ethanol (95%)	70.0 mL
Distilled or deionized water	29.0 mL

Brilliant Green

Brilliant green dye	1.0 g
Sodium azide	0.01g
Distilled or deionized water	100.0 mL

2. Acid Fast, Hot Stain Reagents (Ziehl-Neelson)

Carbolfuchsin

Basic fuchsin	0.3 g
Ethanol	10.0 mL
Distilled or deionized water	95.0 mL
Phenol	5.0 mL

1. Dissolve the basic fuchsin in the ethanol.
2. Dissolve the phenol in the water.
3. Combine the solutions and let stand for a few days.
4. Filter before use.

Decolorizer

Ethanol	97.0 mL
HCl (concentrated)	3.0 mL

Methylene Blue Counterstain

Methylene blue chloride	0.3 g
Distilled or deionized water	100.0 mL

3. Capsule Stain

Congo Red

Congo red dye	5.0 g
Distilled or deionized water	100.0 mL

Manevals Stain

Phenol (5% aqueous solution)	30.0 mL
Acetic acid, glacial (20% aqueous solution)	10.0 mL
Ferric chloride (30% aqueous solution)	4.0 mL
Acid fuchsin (1% aqueous solution)	2.0 mL

4. Gram Stain Reagents

Gram Crystal Violet (Modified Hucker's)

Solution A

Crystal violet dye (90%)	2.0 g
Ethanol (95%)	20.0 mL

Solution B

Ammonium oxalate	0.8 g
Distilled or deionized water	80.0 mL

1. Combine solutions A and B. Store for 24 hours.
2. Filter before use.

Gram Decolorizer

Ethanol (95%)

Gram Iodine

Potassium iodide	2.0 g
Iodine crystals	1.0 g
Distilled or deionized water	300.0 mL

1. Dissolve the potassium iodide in the water *first*.
2. Dissolve the iodine crystals in the solution.
3. Store in an amber bottle.

Gram Safranin

Safranin O	.25 g
Ethanol (95%)	10.0 mL
Distilled or deionized water	100.0 mL

1. Dissolve the safranin O in the ethanol.
2. Add the water.

5. Negative Stain

Nigrosin

Nigrosin	10.0 g
Distilled or deionized water	100.0 mL

6. Simple Stains

Crystal Violet (See Gram Stain)

Methylene Blue (See Acid Fast, Hot)

Safranin (See Gram Stain)

Carbolfuchsin (See Acid Fast, Hot)

7. Spore Stain

Malachite Green

Malachite green dye	5.0 g
Distilled or deionized water	100.0 mL

Safranin (See Gram Stain)

8. Vogel-Bonner Salts (50x)

Magnesium sulfate	10.0 g
Citric acid	100.0 g
Dipotassium phosphate	500.0 g
Monosodium ammonium phosphate	175.0 g
Distilled or deionized water	<1.0 liter*

*Total solution volume is 1.0 liter (The solid ingredients raise the solute level significantly in this preparation. Use only enough water initially to dissolve the ingredients then add enough to bring it up to one liter.)

References

Eisenstadt, Bruce C. Carlton, and Barbara J. Brown. 1994. Pages 21 in *Methods for General and Molecular Bacteriology*, edited by Philipp Gerhardt, R. G. E. Murray, Willis A. Wood and Noel R. Krieg, American Society for Microbiology, Washington, DC.

Power, David A. and Peggy J. McCuen. 1988. Page 4 in *Manual of BBL® Products and Laboratory Procedures, 6th Ed.* Becton Dickinson Microbiology Systems, Cockeysville, MD.

DIFCO Laboratories. December 1995. Package insert L-PI1464-2. DIFCO Laboratories, Detroit, MI.